George Bruce

Shipbuilding Management

Springer

George Bruce
Formerly Professor of Ship Repair
and Conversion
Newcastle upon Tyne, UK

ISBN 978-981-15-8974-4 ISBN 978-981-15-8975-1 (eBook)
https://doi.org/10.1007/978-981-15-8975-1

This Springer imprint is published by the registered company Springer Nature Singapore Pte Ltd.
The registered company address is: 152 Beach Road, #21-01/04 Gateway East, Singapore 189721, Singapore

Introduction

Business has only two functions—marketing and innovation.

—Milan Kundera

In the 1960s, I spent one year as a shipyard apprentice, prior to studying naval architecture. At the end of that year, I was sure that I knew a lot about the industry. Fifty years later having been lucky enough to visit hundreds of shipyards all over the world, meet many great people and work on some remarkable projects, I know the sheer scale, complexity and extent of shipbuilding. The industry has changed dramatically over the period of my career, and yet at the same time it remains very familiar. The shipyards look much the same as the building docks, large cranes, workshops and the major equipment have not changed substantially. The ship is still constructed in much the same way, although some of the ship structures and equipment are very different. Many activities that used to be carried out in shipyards are now sub-contracted, but the industry still has to provide all the necessary ship fittings and equipment.

Of course, both the ships and the shipyards now have access to much more developed technology. Information technology, electronics and electrical equipment in general have offered more capability, but made demands on the management. Welding is another good example where the shipbuilding processes in 2020 are far advanced from those in 1970, especially for large ships with heavy structures, and also in terms of the speed and quality of the process. There are far fewer people working in a typical shipyard, as improved organisation and mechanisation have replaced most manual labour.

Nevertheless, the really big changes in shipbuilding have been in how the industry is managed, how the physical assets and the people are organised. The ability of shipyard managements has been enhanced massively, especially by the ubiquity of computers. Despite this, the current organisation and management methods began to develop more than fifty years ago, and major changes were managed a long time before the IT revolution. If the industry is studied carefully, many of the underlying challenges and their solutions have not changed in principle. So, there is a mixture of continuity and change in ship construction.

In fact some of the basic issues in shipbuilding have not really altered since ship construction began. The requirements are to identify and prepare a suitable site, secure the right equipment, find or train a workforce, motivate them, organise the materials to the site and then build as efficiently as possible. And above all, the objective is to win shipbuilding orders and then make a respectable profit on the activity. The supply chain is now much more complex than in earlier centuries, but obtaining the materials sometimes from international sources was often a problem hundreds of years ago. There was a need to find and secure up to two thousand suitable trees, transport them and then shape parts to fit the complex shape of the ship hull. Unless the ship was built beside an unexploited forest, the business required a major logistical operation.

This book is a personal view of the management of ship construction. There is only a brief discussion of the hard technology, because this is generally well understood. How to organise and use the technology efficiently is the key to success. Some of the ideas in the book are old and often do not rely on current technology. While recognising that the technology is very helpful, I believe that the underlying ideas are more important. After all, Archimedes' principle is still valid after more than two thousand years.

While the book represents my personal views, I have to acknowledge the help of numerous people throughout my career. I have learned from everyone connected to the shipbuilding industry that I have met and worked with. Many I only met briefly a long time ago so sadly will remain anonymous, many have long retired and some are no longer with us, but I gratefully acknowledge their help. Some from my early career stand out for me.

Three staff from Newcastle University come first, Michael Chilton who admitted me as a student despite my examination results, Jim Teasdale who inspired me to concentrate on building ships rather than designing them and Arnold Emerson, my personal tutor, who advised me to find something interesting and unusual when I graduated so I would not be lost in a large technical department. I was then able to join the development group at Swan Hunter during some major shipyard developments and Fred Oman, who ran the project team was an excellent mentor, who gave me some early responsibility as well as lot of practical knowledge.

When I moved on to A & P Appledore (APA), I joined a professional and motivated company which was also tremendous fun and I learned from everyone. I also worked with some specialist civil engineers who were always happy to share knowledge. At British Shipbuilders, the director of the research team where I worked on ship production, George Snaith, introduced me to some novel and challenging ideas. Colin Gallagher who set up the MBA programme at Newcastle University gave me three years of insight into management. This was invaluable when I ran my own company and applied what I had learned to managing a business. Finally, I returned to Newcastle University to teach shipbuilding production and management. There Bill Hills advised and helped me build a research team. Through him I met and formed a good working relationship with Nick Granger, then director of the UK Shipbuilders Association, on programmes of UK and European research.

I am grateful to all these and many, many others. Above all to my wife Louise who has supported me over many years, suffering my late working hours, short notice overseas visits and frequent jet lag.

Contents

About the Author

George Bruce After a shipyard apprenticeship, during which he was awarded a BSc in Naval Architecture, his early career was in shipyard development. Three years were spent in shipbuilding research and development at British Shipbuilders. He worked for a major consultancy internationally, working with private and public shipbuilders and advising governments, and rose to Technical Director. During this period, he was awarded an MBA from Newcastle University. He then ran his own company, consulting and providing management software for ship repair. He joined Newcastle University, teaching marine production and management, and was awarded a chair as Professor of Ship Repair and Conversion, retiring in 2012. He remains active in teaching and consultancy in shipbuilding and repairing.

Chapter 1
Background to the Industry

Without shipping, half the world would starve and the other half would freeze!
—International Chamber of Shipping
Marketing is not anyone's job... It's everyone's job!
—Jack Welch

The starting point for any shipbuilding project is a potential contract. So the market is an appropriate place to begin a discussion of shipbuilding management. Any shipyard can only operate in the context of the market demand for its products. The long term growth of the marine market over the last 100 years, in terms of the annual production of ships is well known. There have though been remarkable peaks and troughs in that growth alongside the continuous trend. Recent shipbuilding in the World has produced around 100 million gross tonnes annually.

Considering only ships of 1000 gross tonnes or larger, there are over 50,000 merchant ships trading internationally, carrying every kind of cargo. The total world cargo capacity is approaching 2 billion gross tonnes and at the time of writing these ships move around 90% of world trade. Although high value and perishable cargoes travel by air, the volume of world trade would be impossible without the large fleet of commercial shipping. Without shipping, the bulk transport of raw materials, of manufactured goods and of food would simply not be possible. Ships are complex, technically advanced and expensive artefacts, with the largest cargo carriers costing over US $200 million to build, and passenger ships up to US$ 1000 million.

Seaborne trade has continued to expand, because of the very competitive freight rates it offers. The cost of transporting a tonne of ore or a container of electrical goods is low, so the cost added to individual items is minute. Because of the growing efficiency of shipping as a means of transport and increased internationalisation, the long term prospects for the industry's further growth continue to be strong, whatever short term issues arise. Overall marine production includes the several industry sectors, although shipbuilding is the focus of this book. Other sectors include ship conversion, ship maintenance and repair, small service and leisure craft and offshore structures.

The demand for ships comes in three broad categories, which are movement of goods and commodities by sea, recovery of resources and use of the sea for leisure activities. These demands require a supply of ships, which have to be designed and

G. Bruce, *Shipbuilding Management*,
https://doi.org/10.1007/978-981-15-8975-1_1

built in the first instance, maintained during their working life, refitted to extend that working life then finally disposed of when no longer operable. For the purposes of this book only a narrow range of relatively large commercial cargo carriers will be considered, although the principles of management described can be applied to most types of ship and also to smaller vessel.

The main types of commercial carrier are container ships which carry most of the world's manufactured goods and products, usually through scheduled liner services, bulk carriers which transport raw materials such as iron ore, coal and grain and tankers which transport crude oil, chemicals and petroleum products.

In addition there are ferries which generally make short journeys with a mixture of passengers, cars and commercial vehicles. Most of these ships are Ro-Ro (roll on-roll off) ferries, where vehicles can drive straight on and off, making the turn round very efficient and requiring only a short time in port.

Ships were the usual and often only means for people to travel long distances until the introduction of large jet aeroplanes. Most general cargo ships on regular routes carried a few passengers, and many of the larger ships carried hundreds. When air transport took away the passenger market, the passenger shipping companies turned to cruising. Demand for cruise ships expanded rapidly during the 1980s, leading to a new generation of what can be described as large and luxurious floating resorts.

There are also numerous specialist ships such as fishing vessels, tugs, research vessels, offshore support vessels for oil production, and increasingly for wind power generators.

The management of any shipbuilding company must begin their search for new contracts by reviewing the market demand for their products. To do so it will often start with some model of market demand. Some shipyards have local markets with limited international competition, others have very specialised products, such as passenger ships, again with limited competition. However the majority have to compete in a difficult international market, so need to understand where their contracts may come from.

A number of factors can be identified which influence the demand for marine vehicles. In the case of the shipbuilding market, these are usually accepted to be first the replacement of ships which must be disposed of because of their age. In general ships last around twenty five years, before age, though sometimes obsolescence, makes them no longer economic to operate. Some ships last longer, such as passenger ships which are frequently upgraded and can last for forty years. Others, for example the last generation of break bulk cargo liners, are used for a much shorter time in this case made obsolete by the introduction of containerised cargoes.

The shipyard must operate in the existing economic environment. Essentially ships are built when there is demand for shipping, which is dependent in turn on the state of the World economy, which drives international trade. Increases in trade have resulted in the construction of the very large container ships, following the earlier pattern for oil tankers, some bulk carriers and, on a smaller scale, gas carriers. The twenty-first century has seen further increases in global trade and a surge in ship construction, inevitably to be followed by another fall.

Predicting the shape of the future economy is difficult, and this book is not in the business of prediction. While the long term growth trend is clear over the last hundred years, because of the massive peaks and troughs in the growth curve this is not of great value to a shipbuilding company. In economic terms that company is looking at the short term, essentially seeking to answer the question of where the next shipbuilding orders will come from.

When demand for shipping is high, a significant factor in demand for new ships is the construction of additional ships to support growth in world trade. At other times there is very limited demand and this uncertainty creates a problem for a shipyard management. Economic data is also useful in that it can support a future shipyard development, or suggest that the time is not right for such a move. It can also provide evidence for seeking government support in difficult times.

Further additional ships are required to support changes in trade patterns, as examples the move to larger container ships for round the world services, which call at very few ports. These are supported by smaller feeder container ships to distribute cargoes to smaller ports. There have also been changes in the technology in ports for loading and unloading ships, to speed the process and keep ships at sea, earning money.

There are also new cargoes and new routes, of which liquified natural gas is an example. This has developed from a small trade to a major export business from the Middle East and other regions with suppliers often owning and operating the ships.

The short term demand for ships is affected by freight rates. As shipping is a very elastic market, a small deficit in the number of ships available for a specific trade, compared to the demand from shippers can lead to very large rises in the rates offered. This tends to drive up ship values and create new demand, or for major conversions if the demand is very short term.

In assessing the market, a company must also take into account the supply of shipyard capacity, part of which makes up their competition. The supply side of the shipbuilding market, and other marine markets, is difficult to assess. There is no universally accepted, or particularly accurate measure for the capacity of the world's shipyards. Compensated gross tonnes is often used, as it can be applied to different ship types, but there are other factors which affect the actual capacity, such as labour force availability, political influences and technological changes.

The supply of shipbuilding capacity depends on the capabilities of shipbuilding facilities around the world, the output capacities of the shipbuilding facilities which are capable and the current availability of those facilities.

The commercial function in a large shipyard may carry out its own research into the market, or may commission a specialist research company to work on its behalf. The outcome will be a forecast for the demand for ships which the shipyard is best able to build. It is possible to forecast demand in general terms, but not so easily for specific products in what is a large, made-to-order business. The market studies will also examine the product life cycles for the ships, because knowing that demand for current products is likely to disappear may also have to be taken into account. Alternative scenarios may need to be considered for future production.

Specific potential contracts will always require direct approaches to potential customers. To assist in approaching shipping companies, using the information gained from a market study, the technical design function can prepare outline designs for potential products. These may be new products, incorporating radical developments or may be current products with small updates.

There is also competition to manage. Even companies which dominate niche markets may have problems with the possible exception of builders of cruise ships and military ships.

Having briefly considered the shipbuilding market, it is worthwhile now to offer a brief review of major changes in shipbuilding over the last 200 years. It includes the change from wood to iron as the main structural material, then the development of steel ships, increases in size and in the variety of ships. Changes in propulsion include the move from sail to steam, motor and more recently electric ships. There is also an increased complexity in equipment fitted aboard the ships which has affected how the industry operates.

Where and when shipbuilding started is uncertain. However, Pacific islands, Australia and New Zealand were populated around 50,000 years ago and given the distances from other continents it is reasonable to suppose that substantial boats or rafts were being created to carry groups and their belongings.

The Mediterranean was an early centre of human development and therefore trade. Egyptian vessels made of wood have been traced to around 4000 BCE. They are also known to have built boats of papyrus bundles. Early trade tended to stay close to coasts, because the vessels available were not strong enough for deep water travel, navigation was difficult and so the crew (and the vessel if necessary) could spend nights on shore. A number of archaeological finds have been made of ships from early Mediterranean trading ports, which demonstrate shipbuilding techniques in wood.

China had early shipbuilders, as did Malaysia and other Asian communities, and trade in various regions of the World was well developed by around 1000 BCE. Ships were relatively small, and fragile, mainly made from wood and powered by sail or oars. Deep water voyages were still difficult and so trade mainly remained coastal or in sheltered waters. The small vessels could also trade along many rivers.

Despite their limitations, early ships were very important to permit large scale trade in eras of limited roads which made overland trade in any quantity difficult and slow. Even a small ship can carry many tonnes of cargo, equivalent to tens of carts or hundreds of pack animals.

In the first millennium CE, Chinese voyages in larger vessels are recorded, and other Asian nations are known to have traded considerable distances. Some Europeans were trading into and probably across the Atlantic before the generally accepted dates of discovery. China invented the rudder, improving sailing ship capabilities and larger ships were built world wide. In the second millennium, major trades were established across the globe, and explorers began to link the different regions establishing genuine, global seaborne trade.

Until 1800 there were modest improvements in ships in terms of capability and size, but the construction from wood generally imposed limitations with a maximum convenient length of around 60–70 m. Sail power limited the ability to trade because ships could only sail with the wind behind them or to a limited extent from the side.

To construct a ship required a sheltered site adjacent to water with adequate depth, a slope so the completed hull could slide into the water, access to large quantities of timber and a workforce. Of the workforce only a small number needed very specialised skills, primarily in developing the shapes of parts and fitting them together. Given the slow speed of construction, and the manual labour involved, a few skilled shipwrights could direct the work. Ships were built by identifying suitable trees, often using curved branches to allow strong frames with the wood grain along their length. These were then cut, taken to the construction site, seasoned, cut to shape and then fitted piece by piece to the hull. Masts and spars were made as far as possible from single, tall, straight trees, often imported to Europe from North America or the Baltic. The few metal fittings could usually be made locally.

The abundance of wood on the North American East coast, alongside a growing population of some 5 million (compared to 10 million in Britain and 5 million in Ireland) made the UK colonies in North Americaan an ideal location. The Atlantic trade created a large demand for ships and along with the UK itself, the New England shipbuilders dominated Atlantic shipbuilding. This continued into the nineteenth century.

The industrial revolution, among other developments, produced steam power and the ability to produce iron in significant quantities. Early proposals were made for steam powered vessels in the late eighteenth century, in the UK, France and the USA. The first quarter of the nineteenth century saw some practical examples of paddle steamers. By 1850, propeller driven steam ships were employed on trades across the world, with building led by the UK as the first major industrial centre.

From 1850, steel production was improved and so steel gradually replaced iron as the major shipbuilding material. The basic construction process did not significantly change during this period. Most commercial ships were still relatively small, and were built up from the keel, piece by piece, in the same way as timber ships. Engines were required, and most shipyards built their own, along with the pumps, winches and other equipment. A ship plate weighing one or two tonnes was the largest item to be lifted, and the work remained dependent on a lot of unskilled manpower.

It is worth pointing out at this stage that there are always exceptions to the typical situation in ship construction. From the Great Eastern of I K Brunel in 1854, through major warships of the late nineteenth century and large passenger liners, particularly on the trans-Atlantic trade, some ships have required different approaches to construction. Large ships and major items of equipment have needed heavy lifting capability as an example.

Military ships have often differed in their construction and management. Although most ships were built on inclines to simplify launching, warships have sometimes been constructed in dry docks, since ancient times. Major arsenals existed in ancient Greece, Rome, mediaeval Venice, Genoa, and other cities. An example can still be found in Barcelona, where the maritime museum is housed in the mediaeval naval shipyard. There is a large, by the standards of the fourteenth century, dry dock and various covered workshops and stores.

The start of the twentieth century saw the beginnings of modern shipbuilding. Although most commercial ships remained small, generally below 100 metres in length, some very large military and passenger ships were being constructed. Size and military imperatives made faster construction desirable and larger cranes, still tiny by current standards were introduced. There was an expansion in trade, and therefore a corresponding increase in ship construction. The UK, which had become the dominant shipbuilding nation in the nineteenth century continued in this role. In some years 80% of merchant ships in the world were built in the UK. However, most European countries were rapidly industrialising, notably Germany, France, Italy and Spain and in the Far East Japan was creating a shipbuilding industry.

The new entrants to the business, without decades of traditional methods behind them, were quicker to adopt new methods. Then industrialised shipbuilding for war time needs was developed in the USA. High rates of production were largely achieved mainly by using many building slipways. There were also some examples of use of the mass production methods being found in other industries such as motor car manufacture.

The 1930s saw a major depression once wartime losses had been replaced and led to many shipyard closures. The shipbuilding industries in most cases in Europe and the USA suffered. The UK in particular saw many shipyards closed. In Germany and Japan, shipbuilding was partly protected as an important national asset. Protection for national industries has been common throughout history. Monarchs have routinely awarded monopolies of specific trades to individuals and companies, and more democratic governments have followed in more recent times.

Germany also made significant developments in the early use of electric welding. The first welded ships were built over one hundred years ago but early techniques were limited and welds were not always reliable. Even when welding became the favoured joining method for steel ships, it was common to rivet the bilge plates and other areas seen as crucial to ship safety.

Early diesel engines began to replace steam at a similar time, but the complete replacement for commercial shipping took some fifty years.

A significant change, albeit one very localised, took place in the USA in the 1940s. The need to build large numbers of merchant ships, to replace losses, led to the liberty ships. These were a basic general cargo ship, constructed to a standard design created in the UK and mainly built in the USA. Many of the sites used were new to shipbuilding, and production engineers played a major role in developing the processes. The biggest change was to construct the ships from units, rather than piece by piece. This massively reduced the construction time and in the longer term set the pattern for most future shipbuilding.

Since the 1940s, there has been steady progress in the development of the shipbuilding industry, led by the increasing sizes of oil tankers, followed by bulk carriers and container ships. Passenger ships, which were the largest ship type from the late nineteenth century, are the most recent to follow the trend of further size increase. Size dictated the need for large steel workshops, large docks and cranes. There was then a requirement for faster outfitting as ships have become more complex and simply to keep pace with the speed of hull construction.

Periods of expansion of the world-wide industry, generally followed by downturns in the market have required shipyards to concentrate on improving productivity, which has been supported by the rapid development of computers and the use of these for design. Some computers were to be found in shipyards in the 1960s, for some design and production activities, but mainly for office functions especially payrolls. This has expanded massively into management information systems to support all aspects of the shipyard operations.

The developments since 1945 can be very roughly divided into phases of technological change. The first phase, up to around 1950 continued with earlier practice, with ships built on slipways, directly from piece parts so the ship was in a fixed position. Much of the outfitting was completed after launch of the hull.

The second phase from 1950 to around 1965 saw two major changes. First the increase in ship size, especially oil tankers, demanded large docks to accommodate them with enhanced steel workshops and large cranes to lift units and blocks. The Burmeister and Wain shipyard in Denmark created a new shipyard with a dry dock and large portal crane which was the pattern for most subsequent shipyards.

The larger blocks were needed to avoid an excessive time to construct the hulls. To support the increase in the volume of steel used, there was also mechanisation of particularly the initial production processes, cutting parts and sub-assembly. Japan, partly by investing in new shipyards but also by adopting good management practices, became the largest shipbuilding nation.

The third phase from 1965 to 1980 started with a large expansion in demand for ships, and corresponding increase in shipyards, especially for larger ships. Further expansion of shipyards followed, including the start of the South Korean industry as a major component of that nation's industrialisation. There was further mechanisation of the steel-working activities, for example the panel line was developed for the larger ships being built. Outfitting was increasingly completed before the ship was afloat, and delivery shortly afterwards became usual. Some vary advanced facilities were developed, in Scandinavia and Japan in particular. A few shipyards were largely or completely undercover, generally for smaller ships.

At the mid-point of the phase, a massive increase in the price of crude oil brought much shipping trade to a halt and shipbuilding entered a recession. Many contracts were cancelled, although some expansion of capacity in the world did continue.

The fourth phase, from approximately 1980 to 1995 saw a lot of consolidation in the industry. Some countries completely or largely abandoned shipbuilding and those that did not were forced to provide large subsidies. Others, notably China, continued to expand. There was a major emphasis on productivity and efficiency and developments in technology focussed on the management. This included significant investments in computer technology for design and production control.

The fifth phase from 1995 to 2010 saw further new shipyard developments, largely in anticipation of a new expansion of demand which was forecast in several countries. The use of computers continued to expand, with greater capabilities. New facilities used the same physical plant and equipment as for the preceding decades, with incremental improvements in specific technologies available.

Since 2010, the size of the world fleet of commercial ships has reached its largest ever. Very large container ships have been built, to join the tankers and bulk carriers, along with many new passenger cruise ships. Technology in terms of the integration of design and production has developed further. Emphasis has been on efficiency of production.

South Korea has continued as the leading shipbuilding nation, closely followed by China. The Korean industry is very efficient at constructing ships. China is behind in terms of efficiency but has the benefit of lower wages so far.

Estimates of world shipbuilding capacity vary, but AWES figures are reasonably reliable. Using millions of CGT as the measure was over twenty million in 1975 and 2000. In the mid 1980s capacity fell to fifteen million CGT, but there was still a surplus capacity of nearly 30%. In 2020 the World can build around thirty million CGT annually.

How the international shipbuilding industry will develop is the subject of many studies, and it is not the intention with this book to add to them. Rather the focus will be on how best to manage an individual shipbuilding company, to survive and prosper against whatever background occurs.

Chapter 2
An Activity Map of the Shipbuilding Process

So much to do, so little time
—Simon Bolivar

Having briefly looked at the shipbuilding industry, it is now useful to outline the process of building a ship and the activities required from start to finish. This outline uses an "Activity Map", which provides an overview of ship production, showing the stages of a ship project and the also major functions of a shipyard which are needed to realise the project. Each intersection of a stage and a function indicates an activity. The main activities will then be discussed in greater depth in subsequent chapters, as will the stages of a contract.

The scope of ship construction management includes every stage of a contract, and pre-contract stages, for all the necessary shipyard functions. The shipyard functions are organised by discipline, and provide a basis for the management structure which keeps the shipyard operational. They are described here as,

Planning, including all stages from initial design to final delivery of the ship.
Design of the ship from initial product development, design for operation, design for production and finally sea trials.
Production of the ship including all production activities and essential support activities.
Production engineering, developing the production methods and ensuring the design and production are in synchronisation to make the production activities efficient.
Quality management, to ensure the product meets all customer and regulatory requirements as efficiently as possible.
Purchasing and materials management, from specification to assembly and installation of bought in items.
Commercial activities, including marketing of the shipyard and contract management.
Human resources to ensure a supply of trained labour and their safety.

G. Bruce, *Shipbuilding Management*,
https://doi.org/10.1007/978-981-15-8975-1_2

The key ship production stages follow the development of a shipbuilding contract, beginning with the strategies which the shipyard intends to pursue to obtain contracts. They conclude with the final commissioning of the finished ship and after sales service. They are listed here:

Shipyard strategy, led by product development, how the shipyard intends to construct its products and marketing of the products.
Contract development, to sell the products to interested ship owners.
Product design, developing the ship to meet all regulations and owner's requirements, and design for production to simplify the design.
Material Procurement to ensure all items are available when needed.
Production, which has three stages, first manufacturing of parts for the assemblies which make up the ship, second, assembly of the parts into units and third construction which is the installation of all assemblies, parts and equipment at a final construction site.
Testing, ship trials and commissioning of the ship to ensure conformance with the specification and quality.
Delivery of the ship, ensuring acceptance by the owner and including post delivery support.

There are also maintenance activities which are needed to keep an organisation in operation. They include several activities which are beyond the immediate scope of the contract management but of great importance. The first is financial management, generally to satisfy local financial accounting regulations, to ensure bills and wages are paid and to maintain the company cash flow. Other activities are needed by any organisation, for example maintenance and security.

Information and communication technology is also of significance as it now forms an important element of all the activities undertaken. This is the basis of the management information systems, discussed in a separate chapter. Environmental management is increasingly regulated by national and international rules, and applies to the shipyard operations and the products constructed. Along with health and safety, this is discussed as part of human resource management.

An activity map is one useful means of reviewing the activities in an organisation. Each activity is the intersection between a stage in the life of a contract and the contribution of each shipyard function to that stage. So the map shows all the main activities and their overall outputs for each stage of the development and completion of a shipbuilding project. All the activities from the different contract stages can also be grouped by function, which may correspond to a department, giving a complete overview. The map is a useful starting point for analysis of a particular shipbuilding operation.

The activity map outlines all stages of contracts and all functions within the shipyard. Each stage of work for a contract has several activities, carried out by the different departments of the shipyard and leading to specific stage outputs which depend on the outputs of all the activities.

The main activities are shown in the map in Fig. 2.1.

Consider first the stages of a contract. These can be summarised.

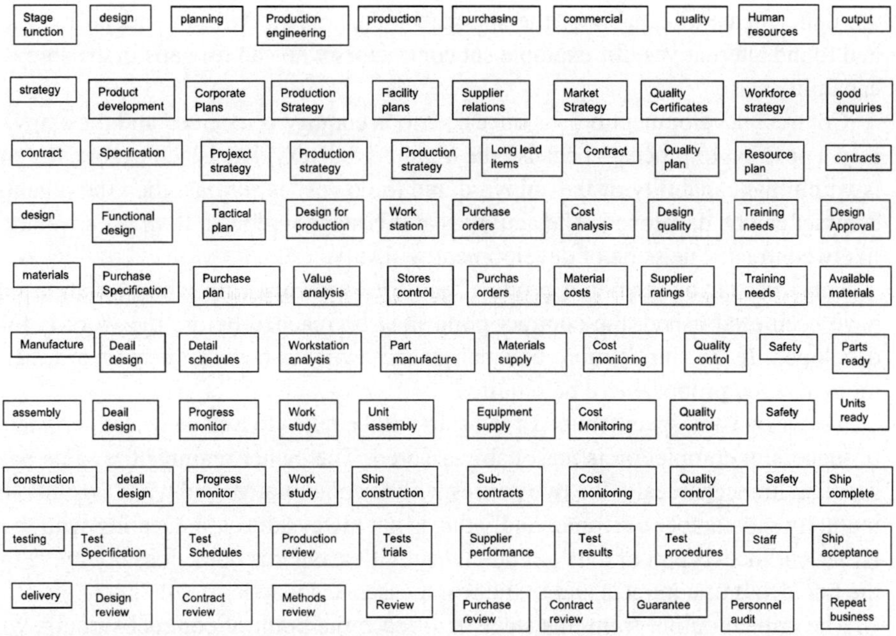

Fig. 2.1 Activity map

Strategic development is essentially a precursor to any contracts, and is intended to position the company to be able to take up opportunities which become available in what is often a changing market for ship construction. This includes ensuring the shipbuilding facilities are upgraded and maintained, along with the overall technology, in a state to support the activities of the commercial and marketing activities.

Product development and marketing is a significant element of the strategic development. How this is carried out by a company depends on particular circumstances. For example a company may be connected to a shipowner who will routinely place orders. Or in the case of military ships, the requirements will be determined by the military and the shipyard only involved at a later stage. Or in some cases such as cruise ships, the product may be developed jointly by the owner and a favoured shipyard which has the track record of constructing similar ships and in whom the owner has confidence.

For most shipyards there will be market research and initial design work for product offerings to ship owners. The output will be a stream of what can be termed "good" enquiries. That is requests from owners who are serious about a ship, financially sound and reliable from the shipyard viewpoint.

Good enquiries will also be for ships which are within the capability of the shipyard, suitable for the facilities which are available and for which the workforce is qualified. As part of the strategy for the shipyard, the quality function will have sets of procedures to ensure the product quality is assured. At the same time, any current or future potential changes in legislation, covering environmental issues, health and safety and working regulations will be incorporated into the strategy. Where there are deficiencies in capability, the shipyard management will have identified these and found alternatives, for example subcontractors who can fill gaps in the shipyard capability.

Contract development begins once a serious enquiry is received and the shipyard is in a position to proceed, because the required delivery date is achievable, the ship is within the capability of the shipyard and the owner is serious, then the potential contract can be developed. The activities at this stage will lead, if all goes well, to a likely contract. The scale of development will vary.

For a reasonably standard product, and for an owner with whom relationships have been established, the contract point may be reached before the work is fully developed. If the ship is more unusual, or the owner is new to the shipyard then a more detailed proposal will be required.

The design and specification of the ship will be taken to a stage where confidence in successful completion is reasonably assured. The owner requires the same basic set of assurances. Ideally the two parties will be collaborative and work together to a mutually satisfactory outcome. Sadly this is not always the case. Dealing with these issues outlined as part of the commercial activities, but this book does not move into the fine detail of contracts and in the worst cases, legal actions.

The critical output from this stage is to reach the point of contract signing, with the shipyard and the owner both confident that the ship can be built for the price and within the timescale asked for. The shipyard should also be confident that the quality required can be achieved, that is the ship will perform as specified.

Product design may already be complete, whether the work is done by the shipyard or, for example, where the design has been sub contracted by the owner to a third party. In other cases where a close relationship has been established, the design may be largely developed after the contract is agreed. At this stage the objective is to take the design to the point where the owner's and all regulatory interests are satisfied.

The owner requires a ship that will perform as required. The regulators are largely concerned with safety, for the ship, its crew and the environment in which it will operate. The functional design of the ship is now completed, if this has not already been done, to ensure performance will be as required.

As the functional design is created, then production engineering is required to ensure the lowest possible cost of production. The way in which the ship will be built has to be a consideration from the start, so that use of available facilities, such as cranes, can be optimised. The block breakdown for the ship and outfitting policy have to be decided so that these can guide the design as far as is feasible. Production engineering is also required if, for example the facilities have to be adapted for a new type of ship to be built.

Alongside the design is the specification for the ship, now completed in detail. As the project develops, the design of more detailed assemblies and parts has to be considered, again with the intention of minimising production costs. As far as is possible, the design details will use standard items, especially where similar ships are produced in numbers. This is beneficial for production efficiency as well as reducing the design work content.

Material procurement is an important stage, because materials purchased from third parties form an increasing proportion of the content of a new ship. The subject is therefore one which is of great importance for a shipbuilding project. As with other functions of the shipbuilding company, the engagement of purchasing should be through the complete life of the project. Pre-contract the purchasing function will maintain contact with suppliers, to understand the current market situation. As design concepts are developed, the purchasing function will begin to consider the supply of long lead items.

At the preliminary stages the focus is on cost estimating, where the purchasing function has a major role. As the point of creating a contract approaches, there is an increased focus on establishing lead times for major equipment items.

Once the actual contract work starts, the focus is on ensuring all items to be bought in are ordered on time and on tracking them. Where there are potential difficulties, the function is concerned with expediting the items. And of course behind all purchasing activities is the wish to secure the best prices possible to reduce the overall cost of the ship.

In the case of a ship where the owner wishes to specify particular suppliers and equipment, those responsible for purchasing will have little control over prices, and may need to pay more attention to delivery expediting. Any risks will have to be identified as early as possible so they can be input to the contract negotiations and planning.

Production effectively begins at the early stages of a project when the production function is concerned with facilities and has an input into the production engineering. There is also a major input into the strategy for construction and outfitting, based on past experience. Once contract work begins, production focuses on managing materials in the shipyard, as these are delivered, then part production, assembly and finally construction.

Part production, alongside procurement from suppliers, creates all the elements of a ship. Major sets of parts are steel plates and profiles and pipes. These are then managed in accordance with the work planning to ensure timely delivery to work stations.

Assembly joins the parts to make sub assemblies then larger assemblies. In some cases relatively small units are assembled to be taken to the shipbuilding site for construction. In most shipyards, large units, and blocks are assembled. These are usually partly outfitted before reaching the building sites.

Ship construction uses the assemblies, of whatever size, to build the hull, superstructure and other elements of the complete ship.

At the end of the contract, production will work closely with the technical departments to plan and manage the testing of the ship systems and then the trials.

The testing and commissioning stage of a contract demonstrates that the finished product will meet all the operational requirements of the owner, as well as the regulations imposed by nation states and other parties. Classification societies often undertake the management of regulatory approvals on behalf of governments and international organisations.

In a small shipyard, the technical team responsible for the initial and then detailed design work will undertake the trials, in conjunction with production specialists. A large shipyard is likely to have a dedicated trials team who will move from contract to contract as they near the time of delivery. The scale of trials will depend on the ship, relatively short for a routine cargo ship but complex and lengthy for a sophisticated vessel such as a cruise ship.

To reduce the time required for tests and trials, a rolling programme will be used, with parts of systems, ship spaces and other features tested as the work progresses. This also allows any potential problems to be identified at as early a stage as possible, so that remedial work can be completed before an item is aboard the ship and often difficult to access.

Post delivery support starts when the ship is completed, has undergone trials and is handed over to the owner though that is not the final completion of the contract. It is all but inevitable that there will be problems with the ship. In nearly all cases these are minor, and do not affect the successful operation. However the contract will state that the shipyard has to remedy defects. Typically a one year period is specified, during which the ship will require shipyard attendance. This can be viewed as a problem, unnecessary expense and something to argue about. On the other hand it can be regarded as an opportunity, to cement good relations with the owner. Basically, if problems occur and are dealt with promptly, it may make the owner consider the same shipyard for future contracts.

Shipyard functions are the separate elements of the shipyard, which each have a specific role in the contract. In many shipyards these correspond to specific departments in the shipyard, although there are alternative ways of organising the shipyard activities. The activity map is not necessarily intended as a template for the shipyard management structure. Consider each of the functions in turn.

Planning includes all related activities, starting with the long term intentions of the company, through corporate planning for current and future contracts. It then moves onto individual contracts, starting with an overall plan and continuing into more detail down to local work station scheduling. After production, the planning then focuses on the testing and trials to ensure that the finished ship reaches the standards in the specification.

Design of the ship begins with a concept design, perhaps responding to a tentative owner enquiry or often a shipyard response to the results of market research. In some cases the design is based on a previous contract. An owner may seek another, similar ship, or the shipyard may offer a proven design to the market.

Once there is enough interest from owners to suggest a contract is possible, a preliminary design is prepared. This will generally take the design development to a stage where serious negotiations can be undertaken. Where a shipyard is established in a specific market sector, the design may again be one already proven in service.

As a contract is near and continuing once the contract has been signed, a functional design is undertaken. This will demonstrate that the ship will satisfy the owner requirements and also be acceptable to classification, national and other marine authorities. The precise way in which a particular design develops will depend very much on the circumstances, such as whether the owner is a previous customer, what level of trust exists between the shipyard and the owner and the speed with which the parties wish to progress.

The remaining design stages are focussed on the development and presentation of information for the production of the ship, up to and including trials.

The production function will closely follow the marketing and negotiation stages, and will use the information to review and where necessary plan to modify the facilities and resources which are available. It is essential for efficient production that the ship is designed in such a way that it takes account of such questions as what is the maximum size that can be constructed in the dry dock? Another question is what is the heaviest ship unit or block that can be lifted by the cranes which are available? Also, what dimensions can a block have to fit a building or often a doorway?

There are cases where the answer to such questions is that the current facilities are not sufficient. If a long term change is expected then the answer is to develop new facilities, or of course simply to decline the contract. If the requirement is short term, perhaps for only one or two ships, then there are often opportunities for improvisation. Heavy lifting equipment can be hired, sections of the ship can be fitted after the ship has left a dock and temporary buildings can be erected. Some ingenious ways have been found to overcome problems, but it must be realised that while the technical challenges can be overcome, there may be a serious additional production cost.

Production will work closely with the design as it develops and ensure the practicality of the design for producibility and also of the plans and schedules as they develop. Realistically, the production team in any shipyard are fully engaged with the day to day work of completing contracts. Once an issue is raised the necessary discussions with designers and development of solutions are generally left to a specific team of production engineers.

Production then undertakes to construct the ship as specified before handing over to test and trials teams for delivery.

Production engineering is a function which works closely with the production and design functions to develop the facilities, methods and equipment needed for the shipbuilding programme. The function also links with design to review and if possible modify designs to meet production needs. Production engineering also generates data for estimating and production where novel design features are used on a ship and there is insufficient previous experience.

This function takes part in the strategy development, starting with the development of how new ship types will be produced. If this requires new facilities and equipment then the production engineers will research the equipment market, identify potential additions and calculate the production rates. If new docks or buildings are required, then they will specify the requirements and liaise with specialist civil engineers to create the layout and major construction which is required.

Quality assurance to ensure the ship is fit for purpose is a very important function in any shipyard. Although the numbers specifically engaged in managing quality may be small, they can have a large influence on the efficiency of production, and by extension the cost. The quality personnel also ensure that the finished quality of the ship is in accord with the specification and the contract requirements.

In order for the small quality team to have the required effect, one of their key roles is the generation of quality procedures. That is to create a framework for the operation of all the work stations so that the methods in use will, if correctly followed, ensure that the quality is achieved. Rather than inspecting outputs from the activities of the shipyard, the objective is to have procedures and processes that produce quality as a matter of routine. There is also a large educational element so that the workforce will follow procedures and self-inspect as work progresses. Management also have to be taught to respond to problems openly, with follow up corrective actions.

Material procurement is of great importance for a shipbuilding project. As with other functions of the shipbuilding company, the engagement of purchasing should be through the complete life of the project. Pre-contract the purchasing function will maintain contact with suppliers, to understand the current market situation. If new products or new technologies are within the potential future workload for the shipyard, there will be a need to identify and evaluate new suppliers.

As design concepts are developed, the purchasing function will begin to consider the supply of long lead items.

At the preliminary stages the focus is on cost estimating, where the purchasing function has a major role. As the point of creating a contract approaches, there is an increased focus on establishing lead times for major equipment items. In a busy market, the lead time for some items, for example castings, can be similar to the lead time for the ship, so establishing this and maintaining contact with potential suppliers is important.

Once the actual contract work starts, the focus is on ensuring all items to be bought in are ordered on time and on tracking them. Where there are potential difficulties, the function is concerned with expediting the items. And of course behind all purchasing activities is the wish to secure the best prices possible to reduce the overall cost of the ship.

In the case of a ship where the owner wishes to specify particular suppliers and equipment, those responsible for purchasing will have little control over prices, and may need to pay more attention than usual to delivery expediting. Any risks associated with owner's supply will have to be identified as early as possible so they can be input to the contract negotiations and planning.

The need to purchase items will continue through the life of the contract and may increase towards the delivery and commissioning. Extra items may be needed. At the completion of the ship, an audit will be carried out to identify any residual materials or equipment items. These may be refundable, or be of use in further contracts. Other items may need to be disposed of.

Commercial activities are an important element in managing a ship construction project. They must be closely aligned with the technical activities so the two are mutually supportive. The commercial function is generally responsible for dealing with the customer and for financial elements of a contract. Prior to any contract, there must be a strategy for the shipyard and development of products which are attractive to the market. Market research is fundamental to successful shipbuilding so close contact with the shipowners, operators and often shippers is crucial.

A good contract is an essential component of a ship construction project. The contract is needed because there will inevitably be some remaining risk which cannot be fully eliminated in the design. Because risk remains, decisions are required on who will manage each risk, how variations in scope will be controlled and how disputes will be avoided. In short how will problems which may occur during a contract be anticipated and managed.

The contract structure is important, and it is usual to follow a standard template, though often with local variations negotiated between owner and shipyard. It will set out the programme with approximate timescales to enable a contract plan to be produced and this must be linked to key project plan dates. It is essential to allow time to check that the shipyard offer is consistent with the customer's requirements both technically and commercially. If differences exist the contract should include agreement on how they will be resolved.

At the end of the project, after the ship has been delivered, there is an evaluation of the contract which will identify both good and poor performance so that future operations can be improved.

Human resources management supports the technical aspects of managing the construction of a new ship. It is critical to keep in mind that ships are constructed by people, not by machinery. Recruiting and retaining sufficient workers with the right skills is a universal problem for shipbuilders. The human resources function will work closely with others in the shipyard, particularly those responsible for calculating the future resource needs, for known and potential contracts.

There is then a need is to develop a picture of how to deliver the output in terms of resources that the company wants. Starting with the current resources, categorised by age and skill set, the future needs will be determined. As part of this there will be a training and development plan to meet the future needs of the business, whether by direct recruitment of qualified people, training or a mixture of both.

The activity map outlines all stages of contracts and all functions within the shipyard, identifying the main activities which are required. Each activity can be modelled as a process with inputs such as raw materials, parts and information from other activities, both at the same project stage and from earlier in the project. Inputs to the activity may come from internal shipyard sources or from external bodies. The inputs vary according to the process under examination. They include planning of the processes, technical information required for the process to be carried out, materials for the products to be made, or parts for assembly and equipment and tools required for the process.

The activities also have outputs, such as assemblies, delivered to parallel activities within a stage or to later project activities. The process outputs may be physical, for example assemblies, technical such as ship design information or operational data such as planning. Their attributes include their quantity, to check that sufficient products been delivered according to plan and their quality, to see that the products or other outputs conform to the specified requirements. In addition the timeliness, to see that the outputs been delivered on time, waste management, so that the processes in the activity are meeting environmental requirements. Sometimes material or time may have been wasted by process failures.

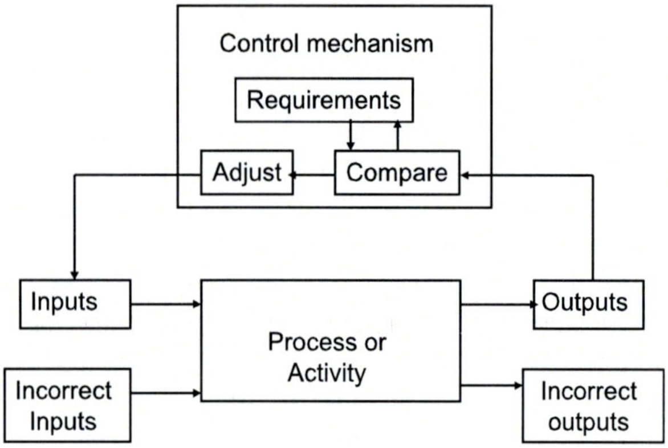

Fig. 2.2 Activity model

All the activities will require resources, which are primarily people who must be adequately trained to carry out the process, and equipment. Equipment, tools, buildings, suitable workshops offering comfortable work environments are needed, with safe and secure storage.

The activities also have constraints, such as performance requirements and timescales to be met. These are aspects which limit the capability and also provide targets for the outputs.

Figure 2.2 shows how any activity can be represented as a black box, that is an initially opaque process. This has inputs which are transformed into outputs, which is what the process is for. It also has a control mechanism to ensure that the outputs meet the requirements set for the process. The control mechanism has a target setting system for these requirements. If the outputs from the process are not what is required, then the inputs can be adjusted. This assumes the process is operating correctly and has been well designed. If this is not the case, then to make improvements, the process itself has to be amended.

It should be noted that not all inputs are necessarily desirable, nor are all the outputs. The inputs and outputs can be planned, as in the case of parts or assemblies, but they may be unplanned as well. Unplanned inputs and outputs are such as poor quality materials, or waste materials, or pollution.

The physical flow of materials and resources is matched by information flows which include customer orders, regulations, standards and work station instructions. Considering outputs such as interim products and unwanted waste material, the corresponding information flows may be such as operating instructions and test certificates.

Performance of the activity is set by the requirements of the overall shipbuilding project. Within this major constraint, which can both limit and stretch the activity, productivity targets are set, which are used by the local management to plan and check progress. Delivery of the outputs on time is required. Cost targets may be set, although it is usual to manage using non-financial targets.

In reviewing the performance of a shipyard, or in planning a new one, the various activities which are needed can be assessed independently. A standard set of questions can be asked. The analysis also applies to the smaller activities embedded within the main ones identified earlier, eventually down to the individual work stations. Questions to ask are about first the technology.

The shipbuilding industry displays characteristics of different stages of development since the 1950s. It is important to be at the appropriate stage of development. It is not necessarily better to be at an advanced stage, if labour costs and production volumes are low. The development of marine production can be expressed in terms of generations, with a significant change occurring approximately every ten year since 1950.

The development of the industry outlined in the previous chapter has touched on how the technology of the industry has changed over fifty or sixty years. This period has seen more change than the rest of history. When assessing an existing shipyard activity or planning a new or improved one, a key question to ask is what level of technology is in use, and should be used?

In some cases, generally for small ships built in small numbers, a very basic technology is used, not dissimilar in many ways from shipbuilding 50 years or more ago. Not only is the technology basic, the performance in both financial and productivity terms is very good. Others using basic technology are less efficient, and the reason is generally the management capability.

At the other end of a scale large shipyards almost all use advanced technology with automation, undercover construction for expensive products where possible and efficient organisation. The initial appearance as shipyard technology improves may not vary much because most of the improvement is in the details of management. The basic shipyard arrangement has not changed significantly for fifty years.

The main issue in making an analysis of the activities in a shipyard is to establish the most appropriate technology to use in order to be competitive. Competitiveness can be defined as the ability to win and execute shipbuilding orders in open competition and stay in business. Ignoring all political and other external influences on markets, what aspects of marine production deliver competitiveness?

Studies have been carried out to use the activity map, or a similar model, to identify shipyard activities which have the most impact on competitive performance. This has been done by comparing the technology (in its widest sense) in use in different activities in shipyards around the world. None of the studies could be considered definitive, and it is difficult to make simple comparisons, but a few general principles can be stated.

First, the need for and effectiveness of the overall strategy and the corporate management are important. Shipyards which have focused on a specific market and all the developments needed to address that market are generally more successful. Having said that, such issues as government support and business structures can outweigh any technological advantage, but considering what is in the direct control of the shipyard, creating and staying with a focused strategy matters. This has been a factor in passenger ship production, in some smaller Dutch shipyards and others around the world.

Second, the commercial activities, particularly marketing and customer care, also have a good effect on competitiveness when managed well. Proactive contacts with the market then rapid and effective responses to specific enquiries are needed. Even though some of the effort spent in this way may seem expensive and wasted, the significance of even a single new contract will outweigh the costs.

Customer care is linked, since a customer who is well looked after following delivery of a new ship is the easiest source of new contracts. Such a customer may also make other shipowners aware of the shipyard responses, leading potentially to more good enquiries. This aspect of shipbuilding has been prominent in South Korean success.

Design of the ships offered to the market is important, but real breakthroughs in design are not very common. Some may result from a focus on a market, for example the X bow for offshore support vessels. In the majority of cases, a solid design to meet owner requirements is needed. Most of the issues are managed through the regulations that must be satisfied and a satisfactory sea trial.

Production of the ship, including all the production activities and their essential support is clearly an important activity. However there is no guarantee that advanced production technology, especially hardware, will offer a competitive performance. A number of shipyard developments have been made over the last fifty years which have failed to perform despite the investments.

Planning of the ship construction, from initial design to final delivery is also very important and was a major factor in the rise of shipbuilding in Japan in the 1940s and 1950s. It may be considered that if it is done well then the ship is likely to be built on time and reasonably efficiently, but this is not guaranteed. However if planning is poor then the result is much less likely to result in a competitive shipyard.

Production engineering, developing the production methods and ensuring that the design and production are in synchronisation can improve the performance of a shipyard, and was a feature of Scandinavian shipyards in the 1960s and 1970s. However, reducing the man-hours required to build a ship has an increasingly limited effect in a high wage economy.

Purchasing and materials management, from specification to assembly and installation of bought in items is, like planning, an activity that has to be done well for a ship to be built efficiently, but does not guarantee competitiveness. If purchasing is done badly, then ship construction becomes very difficult.

Quality management, to ensure the product meets all customer and regulatory requirements is essential. If quality management is poor, then the likely result is a lot of re-work after mistakes are made.

Human resources must ensure a sufficient supply of trained labour. While shortages can be covered by sub-contracting and casual labour, the only long term answer is recruitment and training. Decent working conditions are also needed to ensure retention, because the labour force can be very mobile.

Benchmarking a shipyard's activities against others is a useful activity in seeking to be competitive, and is discussed in the next chapter. It has been done by national shipbuilding associations, and by international organisations including the European

Union. With the reminder that using the latest advanced technology and methods is not necessarily going to lead to increased competitiveness, some useful information can be gained from a benchmarking study.

Chapter 3
Competitiveness

Competition brings out the best in products and the worst in people
—Anon

All shipyards, and the activities they need to perform, operate in a market environment. This is in most respects a competitive market in economic terms, but can often be distorted by political effects. Any shipyard management therefore has to seek to be competitive through improving its business activities, and at the same time manage as far as possible the external environment in which it operates.

The political environment plays an important role in international ship production. In fact, it has been a significant factor in various forms for hundreds of years. Examples include the Venetian empire in mediaeval times banning construction of Venetian ships by foreign builders, but having to relent due to shortages of capacity and skills. In the late nineteenth and early twentieth century, large, transatlantic liners were subsidized by governments for carriage of mail. Most government support involved specialist ships, or military ships, but nations establishing a shipbuilding industry usually provided incentives to their shipowners or builders in general to encourage production in their own shipyards.

During the twentieth century, and into the twenty first, political influence has continued to be a factor, in several areas. The policies of a number of governments around the world have resulted in a degree of support for shipbuilding, sometimes openly but more often in a covert manner. First there has been support for the industry to maintain employment. This was a feature in Europe during the late twentieth century as the industry declined relative to the growth of shipbuilding in Asia.

Second there has been support for shipbuilding as a vehicle for industrial development. The industry is a large user of steel, so supported the development of this basic industry. It is also a relatively large employer of labour, and ship construction requires a significant secondary, support industrial base. Japan rebuilt its general industry after 1945 with shipbuilding as one of the core components, with the ships built often also owned and operated by Japanese companies. This supported the national need to import raw materials and to then export industrial goods to pay for them, to the benefit of the national growth. This was followed in the 1970s and early

G. Bruce, *Shipbuilding Management*,
https://doi.org/10.1007/978-981-15-8975-1_3

1980s by the Republic of Korea, seeking to develop an industrial base. The country followed a similar path, identifying a few key industries, of which motor car and ship production are good examples. More recently China used shipbuilding as part of its general industry strategy and in the process become a major supplier of ships to the world.

Other nations have tried the same path with varying degrees of success, including Vietnam, Philippines and Brazil. A different approach in the USA followed the end of the so-called cold war with the USSR, when military shipbuilding which was the mainstay of the US industry declined. US shipbuilders looked to replace the lost business by building commercial ships. Despite some advantages including relatively low labour costs, this conversion was only partly successful.

In all cases, the political input has supported either the creation or maintenance of the ship production capability of the nation. Much of the current state of development of political influences dates from 1973. The rapid rise in oil prices in that year and subsequent turbulence led to probably more political influence than previously. In previous economic depressions, the 1930s being the major example, politics had dictated whether industries in different countries survived. Similarly in the 1980s, Sweden as a technically very advanced and efficient shipbuilding nation decided that the economics of the business were not appropriate for government support and the industry there went into a very fast decline. The UK government took a similar line in the 1980s and 1990s, as did a number of other European countries. An exception is often military ship construction, and there are almost always local ferries, fishing vessels, offshore support, tugs and other service vessels which do not figure prominently in the international market.

During the 1980s, some governments were providing large subsidies to shipbuilders, because of the low demand for ship construction. European shipbuilders regarded Asian competition as subsidised and unfair, while Korean shipbuilders made the same observations about Europe. The Organisation for Economic Cooperation and Development (OECD) was keen to eliminate subsidies from world trade, including shipbuilding, but there was failure to finalise an agreement on the phase out. Cases taken to the World Trade Organisation (WTO) claiming unfair Korean and EU pricing have also had no effect.

The competition directorate of the European Union issued directives which were intended to formalise the use of subsidies by European governments. They also decided to set standard subsidy rules for Europe in order to promote fair competition.

Their long term objectives were to phase out subsidies entirely to allow a restructuring of the industry with the intention of leaving fewer, but more competitive shipyards. To underpin the directive, the European Commission obtained annual studies to compare European shipbuilding costs and Asian shipbuilding prices. From these studies the market price available from Far East shipyards was found to be as much as thirty percent lower than the quoted cost from European shipyards for the same ship. The differential decreased in the late 1980s and early 1990s, when the EU hoped to phase out any subsidies.

Although the existence of subsidies can be deduced from analysis of shipbuilding output, productivity and international equipment and materials costs, direct proof that they are in use is very difficult. The structures of some large corporations with a shipbuilding division, and of ship equipment manufacturers, can be opaque. Transfer pricing between different parts of the business and selective government support for some parts can easily disguise the subsidy regime. The problem of subsidies whether overt and so potentially amenable to treaty, or covert which makes their existence extremely difficult to prove, is unlikely to go away.

The balance of supply and demand is fundamental to these problems in shipbuilding. There have been serious efforts by some countries to reduce capacity. In employment terms, the total workforce in Western Europe fell between 1975 and 2000 from around 450,000 to 90,000. In the same period in Japan employment fell from 360,000 to 100,000 including sub-contractors. However, over the same period there was expansion of capacity in South Korea, China, Vietnam and other countries.

The most optimistic shipbuilding market forecasts have always assumed that long term demand and supply will move into balance, but this has not happened and when demand has increased so has capacity. The Korean and Chinese demand forecasts during their recent expansions were much more optimistic than those produced in Japan and Europe, where any additional capacity was seen as not only unnecessary for the requirements of shipping but also a threat to their shipbuilders.

In fact there was a significant spike in shipbuilding output, but whether this was driven entirely by increased demand for shipping or by other factors or a combination of factors is open to interpretation. In 2020 there is currently a surplus of capacity over demand for ships, and in the longer term, it is probable that shipbuilding capacity will continue to exceed demand. As a result, continuing political influence is inevitable.

The support measures which various governments have given to shipyards take various forms. These include domestic market protection, investment and research support, direct production assistance, export credit assistance, assistance to customers such as tax breaks for shipowners, home credit schemes and favourable tax treatment. Not all of these are particularly transparent. Both Japan and Korea have had some controls on ship purchases from other countries, and there are cross shareholdings or other financial links between owners, builders and banks.

Whatever the political climate in which a shipyard operates, and whether the influence it can exert on that climate is small or large, the management should seek whatever benefit they can gain. But it remains essential to seek to be as competitive as possible through the management and improvement of the shipyard activities.

For a shipbuilding company, competitiveness can be defined as the ability to win and execute shipbuilding orders in open competition and stay in profitable business. Ignoring all political and other external influences on markets, it is useful to consider which aspects of marine production deliver competitiveness. Various studies of the industry have been carried out to try and compare elements of shipbuilding technology between shipyards to determine which of these elements can improve the competitiveness of a company. Technology was defined so as to include the whole business, not only shipyard hardware. These elements studied were similar to those on the activity map which are listed below.

Strategy and corporate management is the first, overarching stage with the objective of setting overall targets for the company. The first is to set the overall production level, then the required employment levels. This and the annual employment cost can be used as comparisons with others in the market. Potential targets would be to increase output or reduce employment costs. From these it is then possible to calculate the output per employee, and the cost per compensated gross tonne (CGT). Value added per employee is also a potentially useful comparison with other shipyards.

Then the eight main functions found in a shipyard can each be reviewed to determine their influence on competitiveness.

The commercial aspects include marketing and customer care with the main objective of obtaining good enquiries. The number of these received is a measure, though their quality might be subjective. The target would be to have an increasing number of such good enquiries over time. The number of enquiries converted into contracts, and the cost of lost enquiries may also be useful. A further very useful measure is the enquiry conversion ratio in specific market sectors or regions, to help target marketing at the best opportunities.

Purchasing and material control is intended to ensure that all shipbuilding contracts receive the correct materials on time and at an acceptable cost. Useful measures of performance include the number of late deliveries, and the number of rejected items arriving at the shipyard. Measures can also be developed for the supplier performance, which will inform future planning. The costs of materials also provide a useful measure to inform supplier selection, again in absolute terms and in seeking an improving trend.

Human resource management has responsibility for the supply of labour in the shipyard. Useful measures are the rate of labour turnover, the ratio of apprentices and other trainees to the total and the retention rate of those newly trained. It is also appropriate to locate health and safety in this function, measuring the number of reportable accidents and the absenteeism. In both case a falling number is to be sought. Environmental management is also a feature, looking for minimizing waste, costs of disposal and any pollution incidents.

Design and technical capability is the basis for the creation of good ship designs. An acceptable design will be ensured by the acceptance of it by the owner and the regulators. For the shipyard management, measures should also focus on the costs of design, mainly the labour costs and seeking a reduction over time for a given output. Numbers of amendments and any mistakes can also be measured, looking for reducing trends.

Planning and organisation of work will be judged along with production by the completion of work on time. The numbers of work packages, interim products and project milestones which are late are all important measures.

Production engineering works with design and production to make the design outputs as easy to produce as possible, and also to design the production activities to make the interim products as efficiently as possible. The measurements of the outputs are largely subsumed in the production outputs, when they make a contribution to productivity.

Production technology can be measured in terms of the technology which is in use, judged by how up to date it is. However as using the latest technology is no guarantee of performance, it is also essential to measure the actual output. Senior management looks at major events in the life of a project, but at a more local level the emphasis is on completion of all elements of the work on time. In addition the productivity which is achieved is important as a measure of cost control, and the ideal is an improving trend. The extent to which the ship is outfitted at each stage of production is a good measure as a proxy for efficiency.

Production is also responsible for delivering quality output, through the way work is planned and organized. The output will be quality checked but finding a defective item will not avoid the necessary rework, which will affect both productivity and timeliness.

The quality function offers a service to the other functions, first in assisting with the design of operational procedures to obtain consistent and efficient outputs and second by checking those outputs for conformance with the intentions.

The studies which were carried out identified those aspects which were considered to be of most importance in different regions. There was no overall consensus on the most important, but there was some correlation between the competitiveness in terms of cost per tonne of building a ship and the capability of the strategic and corporate management. The key attribute for success was a focused strategy. Marketing was also important, essentially because without successful marketing there are no contracts to fulfill and without contracts there is no business.

Planning in its widest sense also played an important role, once an order was received, but the winning of orders came first. Advanced production technology was no guarantee of competitive performance, nor was design. However in some shipyards purchasing is a major influence on performance, where the shipyard is remote from suppliers or has problems with importing items because of customs regulations.

Human resources was not identified as significant at the time of the studies, for the shipyards which were included. Human resources though is really a major issue for almost all shipbuilding. Basically, it is generally becoming more and more difficult for companies in developed nations to recruit, train and retain skilled labour.

Some aspects of the technology can have a negative effect if they are done badly, which can be as significant as the positive effect of doing them well.

There was a loose correlation between the technologies employed in shipyard and their performance in terms of labour cost per production unit. It could be said that advanced technology would not guarantee good performance, but gave a better opportunity for improvement. Equally, old technology did not necessarily result in poor performance. In some cases of small shipyards, very high performance was achieved with old technology, but very good organisation. For large shipyards, high performance required more advanced technology than for smaller ones, to manage the large scale of production. A tentative conclusion is that organisation and management are at least as significant in achieving good performance as the physical technology in use.

The eight functions can be divided into the activities at each of the stages of a ship production project, as on the activity map. Each of the activities thus identified can be further sub-divided, eventually down to individual work stations or work groups. This provides a useful basis for a more detailed analysis of a shipbuilding company. Analysis is important in reviewing the technology in use and in developing the technologies to make improvements.

As an example, consider production technology. For a detailed analysis of production technology, several further divisions can be identified. Steelwork generally has three stages, unit or block assembly, sub-assembly and part production. Within each of these there are various activities.

For production of blocks these are stages of assembly joining, which is the fitting of several assemblies together and fairing them. Then, after inspection to check the alignment, usually tack welding to hold the assemblies in place. Inspection follows to ensure correct alignment, then welding to join the assemblies and complete the structure. Further inspection, to ensure that the structure has remained dimensionally accurate and the welding is correct is next. Cleaning is required to remove temporary fastenings, debris from the welding processes and any other unwanted material. Positioning and welding any supports for pipe work or other outfitting, fitting pipes or other items, then a final cleaning. The last activity is coating which may be the final coating especially in machinery and cargo spaces.

In the development of the ship production industry, new technologies have emerged in particular since the early 1950s. There have been some large changes often associated with the development of ships, especially size, and the state of the industry in general.

Each activity or element of technology which is identified can be examined, and the technology in use can be at any stage along the path of development. To take the example of steel cutting, this has developed over the sixty years or so.

At the start the usual process was by hand marking of the steel plates using wooden templates and then cutting using a hand cutting torch. This was followed by basic mechanised cutting machine machines. These were simple electric powered carriages with a cutting torch mounted on them. Control could be by running the carriage along a portable metal track, usually for straight cutting.

A significant development was an improved control system, using drawings rather than templates. This used a portal, not unlike a small crane, running on parallel tracks which were permanently mounted on concrete walls. The steel plate was located on a raised support and the portal could move over this, along the tracks. Mounted on the portal were one or more cutting torches. The movement of the portal and torches could be guided by an optical device following a drawing, usually one tenth of full scale.

The optical control was superseded by numerical control, generally using a paper tape with the instructions coded using holes punched in it. Initial development of these cutting machines was in the mid 1960s and early 1970s, coinciding with very large oil tankers, and other ships. The new equipment allowed the cutting quantity to be increased without excessive cost.

The paper tape was replaced by direct control from a saved design, once computer aided drawing became available. Finally as the computer capability improved, direct numerical control from the detailed design was introduced.

The actual steel cutting can also be carried out by different processes, so a fuel gas/oxygen process was used first, which has since often been replaced by a plasma cutting process.

Similar stages of development can be found for all the activities in the industry. So a general idea of six stages of development in marine production technology can be identified. It should be emphasized that there are no fixed dates in this so those below are representative only, nor changes in all activities at each stage. Also, the most advanced shipyards have driven the advances because of an urgent need to improve productivity and replace scarce skilled labour in an effort to stay competitive.

The key features of the technology which define the current stage of shipyard development can be summarized. These are undercover assembly and construction, using large, fully outfitted and coated blocks. There is use of flexible automation and an integrated computer management system which includes CAD, material management, planning and scheduling. Simulation of proposed production is used to evaluate schedules. Close collaboration with important suppliers and good human resource planning, including environmental management and safety. Advanced cutting and welding processes, possibly including lasers in some applications are to be found. Some ships with novel structural arrangements, benefiting from low weld induced distortion.

Using the idea of the development of technology over time, the status of a company can be reviewed. For each element of the technology, in the eight functions, a description of the processes and organisation can be developed for all the activities. The existing technology in the company under review is then assessed, and assigned to the appropriate stage.

This can be done using a simple, numerical scale.

1. This describes the basic technology typical of shipyards in the early 1950s using manual cutting, small cranes, much of the work in the open air, outfitting after lunch of the hull and rudimentary planning.
2. In the next stage there is some mechanisation, typical of shipyards in the early 1960s, some larger cranes for the lifting of units onto the ships and some early outfitting.
3. More mechanisation, with ships constructed from large blocks, typical of 1970s shipyards. Use of panel lines, large cranes, docks not berths, some computers in technical activities.
4. Most work carried out under cover, typical of shipyards of the 1980s. Availability of basic CAD, good planning, early outfitting of most units and blocks.
5. Improvements in much of the technology at a detailed level, with integrated CAD, effective production planning and automation of some early processes. Typical of shipyards up to 2000.
6. Ship factories, with an emphasis on purchasing, integration with suppliers and improved management control systems.

It is important to be at the appropriate stage of development. It is not necessarily better to be at an advanced stage, if labour costs or production volumes are low. As such the scale is not a judgment, and the activities in a single company may display characteristics of most stages of development. It may be appropriate to use level 1 or 2, where the production is for a single ship, where the cost of labour is very low and where time is not an important issue.

Most major shipbuilders are at level 5 or 6 and most are actively looking at further development of technology. Each activity can be given a score using the scale. A mean score for each of the major functional areas can be calculated and this can then be compared with other shipyards. This can be done for individual activities or for the whole shipyard. Some studies have been carried out making such comparisons and they can offer useful insights for a shipyard management.

The technology in use is not a perfect guide to competitiveness, but there is some correlation between the two. High performance in terms of labour hours through advanced technology is no guarantee of financial performance. Although money has been rejected as a usable measure for operational management, it is still important. The ultimate measure for a shipbuilding company is profitability, or at least financial survival.

For each of the activities in the shipyard, some measure of performance is required. The performance measures will allow the shipyard management to assess both the absolute performance and whether that performance is improving over time. If these measures can then be compared with others through benchmarking then an understanding of the relative competitive position can be gained. An overall measure of competitiveness is also useful for comparisons.

A measure of productivity which allows reasonable comparisons between shipbuilding regions, and with sufficient data also comparisons between shipyards is compensated gross tonnes (cgt) produced per man-year. Cgt is a measure developed by the OECD to try to make comparisons, between shipyard regions initially, for shipyards constructing different types and sizes of ships. All ships have tonnage measurements, one of which is gross tonnes (gt). This is essentially a measure of internal volume of a ship, calculated on a standardized basis. It is also partly a function of the ship dimensions.

The OECD in 2007 defined cgt as a unit of measurement designed to offer a common system to compare the shipbuilding output in regions of the World. The compensation factors essentially describe different ship complexity, and hence the work required to construct different ships. There is also account taken of ship size, so a large merchant ship has less work per gross tonne than a small one. In simple terms a bulk carrier is simpler than a passenger ship, and a large bulk carrier is simpler than a small one.

When first developed the measure was based on a applying a compensation factor to ships according to a set of size bands. This led to anomalies where a small difference in tonnage between two ships close to a band boundary could result in a large difference in cgt for basically the same ships. As a result a formula was developed where the cgt is equal to a factor for ship type, to account for complexity, and a function of the gross tonnage to account for size.

It is possible to calculate an accurate total for ship production in a region, usually with a moving average over three years. Then a less accurate but useable total for employment and labour cost can be calculated. The information can often be obtained from published sources, such as national employment statistics, wage statistics, shipyard annual reports and national shipbuilding association reports. Although developed for regional comparisons, for which the method offers reasonable comparisons, the measure has also been used for comparing shipyard groups and even individual shipyards. The more detailed the comparisons, the less reliable they will be.

Problems with the cgt measure include the first relative use of sub-contractors, where the information is buried in bought in equipment and materials costs. So some allowance must be made for the effect of this. Second, even for similar ships there are differences in equipment outfit, in speed and even in quality despite the levelling effects of classification and other regulators. Both of these factors will affect the work content of ships.

Third, where a series of ships is constructed, there will usually be a learning effect so the later ships require fewer man-hours than the earlier ones. This is a factor that will affect the man-hours for a ship. However, with these caveats, the use of cgt is the best available way to make international comparisons, and it is regularly reviewed by OECD. Once the information for a shipyard, and its competitors has been obtained, and corrected as far as possible, a line, or series of lines can be plotted which link companies with equal costs. This can offer a way to benchmark the performance of a company.

A line is plotted for the size and type of ship that the company is constructing or which form part of the potential market strategy. The line links shipyards which have the same cost per cgt. Those with high productivity and high labour costs can be competitive with lower labour cost shipyards which typically have lower productivity. If the actual or planned cgt per man-year for the company is on the line, then it is generally competitive. If the company performance plot is above the line, then it can be considered to be ahead of competitors. On the other hand, if it is below the line then it is not competitive.

This should not be relied on for small variations in performance, but will identify large differences between shipyards. Variations in labour productivity are often large, with the most productive shipyards up to ten times more efficient than the least productive. The variation has been increasing over the last fifty years, as new entrants to shipbuilding begin with very low productivity and take many years to improve in most cases.

Figure 3.1 shows how the cgt measurement can work. The figures used are representative but do not relate to any specific regions or shipyards. There are six points, each representing the cost and productivity for a single shipyard.

The first, shipyard 1, has low labour cost but also low productivity. Shipyard 2 has the same productivity, but lower labour cost. As a result, shipyard 2 has a better performance and is more competitive.

Shipyard 3 has a higher labour cost than the previous cases, and higher productivity. It is still in a similar position to shipyard 1 in terms of competitiveness. Shipyard 4 also has higher productivity, equal to shipyard 3, but lower labour costs make it more competitive.

Shipyard 5 has very high labour costs, but its productivity is lower than required to be competitive so it is in a similar position to shipyards 1 and 3. Shipyard 6 has

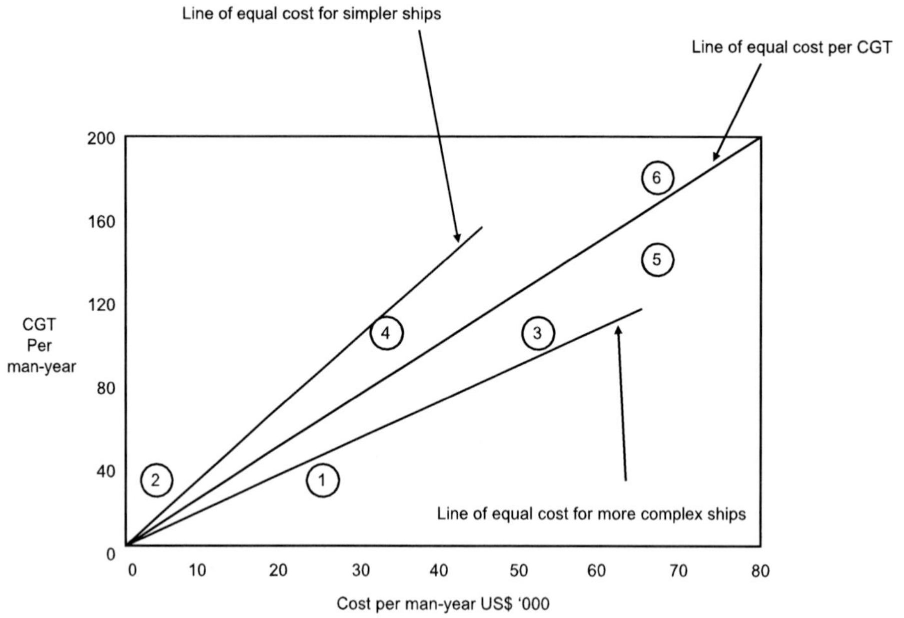

Fig. 3.1 Comparisons of competitiveness

very high productivity, with the same labour cost as shipyard 5, so is able to compete with shipyards 2 and 4.

By carrying out a benchmarking study, a shipyard can determine if there is a need to improve overall performance in cgt per man-year. Incidentally, the effect of subsidies can also be deduced, where a shipyard which seems demonstrably uncompetitive is able to remain in business.

The shipbuilding market is not homogenous, with several major market sectors and a large number of niche markets within the total. For some sectors, different cost structures may be viable due to external factors. As an example, large passenger ships are technically advanced and current builders have developed technology and expertise over many years. This makes the sector difficult to enter and so a different cost of production may be acceptable, because a new entrant to the market would not be competitive. Similarly for military ships, the technology of the product and government restrictions on potential builders, especially other nations, may limit the competition and permit higher prices.

So on the figure there is a second line of equal cost which indicates that a higher cost per cgt may be acceptable where the competition is limited. In this case, shipyard 3 could be competitive if shipyard 4 lacks the capability to build the more complex ships.

The third line shows a lower cost per cgt where very simple ships are to be built, or possibly if the shipyard cost is subsidised. In this case shipyard 2 with the very low costs of labour will improve its competitive position.

Chapter 4
Overview of Project Management

Why do so many professionals say they are project managing,
when what they are actually doing is fire fighting?
—Colin Bentley

This chapter gives an introduction to project management, starting with a look at some main features of a project. So the first issue is to consider what does make a project. Projects are something different from regular production of simple items in large numbers, which characterises much of industry. Basically they are not routine operations. Although many of the activities in ship construction are relatively routine, and often ships are built in a series of more or less identical vessels, each ship can be considered a project.

Other projects often involve change, that is development of a new product, a new operating system or otherwise something that did not previously exist. Change projects cut across departmental boundaries, so can be disturbing for workers and perhaps more for their supervisors. Projects are relatively complex, compared to normal routine operations, partly due to their multi-disciplinary nature.

Projects have a finite life, that is a start and finish date after which the project no longer exists. This is clearly the case for a ship construction project, although the life of the project will usually continue for a year beyond the physical completion and delivery of the ship, where issues under guarantee have to be resolved by the builder. Ship projects include a wide variety of activities and processes, which have to be integrated requiring workers in different technical disciplines to work together.

A project can be defined as a set of linked activities conducted in an organised way, with a defined start and finish date. The activities are intended to achieve a specific result which will meet a defined organisational need.

It is useful to start with the project life-cycle where four phases can be identified. Other versions of the life cycle with more or less phases can be used, although they all follow the same basic life cycle pattern. This is illustrated in Fig. 4.1.

Phase 1 is definition. This encompasses all the activities which are required to create a project and take it to the point where a decision to proceed can be made with reasonable confidence of success.

G. Bruce, *Shipbuilding Management*,
https://doi.org/10.1007/978-981-15-8975-1_4

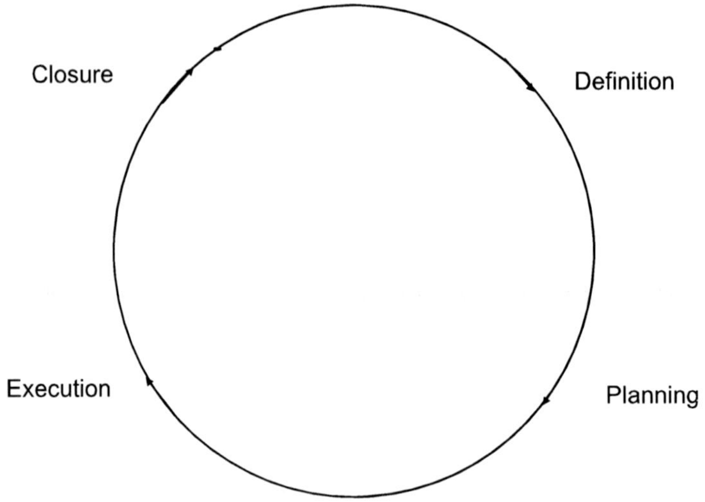

Fig. 4.1 Project life cycle

Phase 2 is planning. This includes all the activities which are required to take the project from a definition of what is to be achieved (for example the creation of a new ship with agreed characteristics) to a definition of how it will be achieved.

Phase 3 is execution. If the definition and planning have been done thoroughly then the project should proceed to plan and reach the required output. Project management is sometimes considered to be about the execution phase, and dealing with problems which arise during execution. This is part of the management, but really the definition and planning are more significant for success for most projects.

Phase 4 is closure. Completing the project requires the project to be carried out as specified, but beyond that the customer, regulators and other authorities must be satisfied. Acceptance of the completion is a significant part of the project as is a thorough review of what went well and less well as a basis for future improvement of the project management process.

The first phase of the project starts with an overall goal. Although the focus of the book is on the building of a ship, the nature of large projects is that they are often part of some larger project. That is, there is a higher outcome than the completion of a specific project, and also many lesser outcomes that build up to that completion. All of these are likely to be projects in their own right. First consider the overall goal which may be different from the project outcome. It should be a change or development which will result in improvement, preferably to the profitability of the organisation for which the project is undertaken. In some cases, for example "blue sky" research, the goal may be more of an aspiration, reachable if the project delivers the outcome which is hoped for.

Considering a ship, the project from the point of view of the shipyard will be to deliver a ship on completion of its construction. This will be the goal for the company contracted to build the ship. However for the owner or manager of the ship, its availability is a means to an end, which is profitable carriage of cargoes. The operation requires one or more customers, whose goal will be to make profits from the sale of goods in one location which have been manufactured, or extracted, or grown, or processed in another. This trade may be encouraged by a government seeking to improve a balance of payments to help their economy, which offers another goal.

Major suppliers to the shipyard will have their own projects within the overall ship project, for example the supply of a main engine or installation of an electrical system. These suppliers then are undertaking sub-projects of the ship construction project.

So the overall goal of a project may be removed from the ship construction. For the shipyard the overall goal will be to remain in profitable business and the successful ship project will be part of achieving that goal. A shipyard may also have other on-going projects, for example for development of new production facilities.

There is often a need to select the best project, or simply to decide whether a project is actually worth pursuing. This will be an issue for a shipowner who wishes to build a new ship or for a shipyard the choice between different development possibilities. All projects require a budget, and setting this is an important step. It is necessary to calculate the expected costs for the project, then the potential income and cost savings. From this a cash flow analysis can be carried out, with some analysis of the sensitivity of the project to changes in external conditions, such as the market for shipping or currency exchange rates.

Making the cost estimates is not a precise science, because there may be unknowns which affect the cost and it is well known that many projects suffer cost overruns Once an acceptable cost estimate has been established, the value of the project can be determined. The value can be in terms of increased output and therefore income, or lower production or operating costs. There may be a benefit from adopting new technologies, although all non financial benefits should really be quantifiable. The estimate of cost for the project may not be accurate if the final costs and benefits differ from those initially estimated and assigned.

The project objectives are what the project should deliver. For example if the goal for a company is administrative cost reduction then the project may be an improved IT system for that administration, or a goal for a port could be a faster materials handling system to reduce ship turn round times. It is important that the goal is not lost in the process of carrying out the project.

The objectives for a project with uncertainty about success may require that the project is split into smaller, phased projects, with interim objectives. In such a case, if the interim objectives are achieved then the project can proceed to a next phase. If the interim objectives are not achieved then there is a decision to be made on whether to continue to try and achieve these before proceeding or simply to stop the project as unlikely to succeed in its overall objectives.

Again considering a ship construction project, then the objective is the successful completion of the ship, for the benefit of the owner who will successfully operate it and the building contractor who will make a profit from the contract. Circumstances may defeat these objectives, for example a sharp downturn in a shipping market may thwart the owner. It may even result in the project being cancelled. Examples such as oil tankers in the late 1970s can be found where ships were never able to trade successfully. Equally, the building contractor needs to complete the project according to a pre-agreed timescale, to an agreed quality and within a cost budget. These are also project objectives and failure to meet them can result in a shipbuilder going out of business.

Once the goal and objectives are clear then the scope of the project can be clearly defined. The scope of the project is a description of what has to be done to achieve the objective. At the start of the project, this may be a relatively limited description. For a ship project this will include such aspects as key dimensions, capacity, speed, power and cargo handling which will achieve the shipowner's goal. As the project develops, and moves from a preliminary design to a contract and then on to regulatory and production requirements, the scope will become more detailed and clearer. Having a clear project scope should remove any ambiguity in what is to be delivered, and thus any potential for disagreements. This should then avoid "project creep", where the scope of work to be done increases especially towards the end of the project, with potentially bad consequences for meeting cost and time objectives in particular.

In practice, standardised contracts, use of published standards from classification societies and regulators and other norms should deal adequately with scope. However for specialised ships there can be elements of the owner requirements and specification which are sufficiently unusual to require careful definition if ambiguity is to be avoided. Such projects are risky for the shipyard undertaking them.

Exclusions from a project may be as important in some cases as what is included. For many projects other than ships there is greater uncertainty about the outcomes. Here it may become important not only to specify what is in the project but also what will not be part of the final delivery. For ships, this aspect is usually built into the standards to be applied and the detailed specification for the marine project. It is important that the contractor for a project is able to control any requested additions at the start or during the project. Even for a ship project, additional work can be a cause of extra cost, delay and also where a contractor has several contracts at the same time potential effects on other projects.

Deliverables are precisely what the client will get. Again clear definition of the scope of work of the project should avoid any ambiguities. For a ship, the specification is critical in providing a precise definition of the deliverables. This includes the standards to be applied, which should prevent arguments about completeness or whether a specific element of the ship is acceptable.

The principle is also valuable for interim deliverables which can be described as any part of the contract which is supplied to the contractor. For example a piece of equipment, large or small from a supplier, or part of the ship which is to be provided by a sub-contractor are interim deliverables. For the supplier or sub-contractor, the particular supply is their project, and the objectives and scope need to be defined in the same way as for the complete ship.

This concept of a deliverable can and should also be applied to parts of the project from another department of the company carrying out the project. In order to achieve the timescale and quality for the overall project, each interim product should be defined in scope so that what is delivered to a subsequent stage of construction is complete. In the past it has often been the case that inaccurate or unfinished assemblies or parts are delivered, for example from the steel workshops to the construction site. Any unfinished work has then to be carried out in less favourable conditions, resulting in rework and serious delays.

Dependencies and constraints may exist for the project. So as well as considering the objectives and scope of a project it is also important to look at any issues external to the project that can have an impact on reaching those objectives There may be other projects competing for resources to complete the work. In a company with several projects it may be considered more important to complete a project close to delivery than to progress one that has only recently started. An obvious reason for this can be the need to secure a completion payment for company cash-flow.

Other projects can complete for labour resources, or for physical resources such as dry docks. A company may be seeking the best outcome from a situation with several projects and one may be sacrificed for the benefit of the whole organisation. Obviously such a situation should not be allowed to occur, but it often does in reality. Decisions of this nature are basically down to commercial judgement and fall outside the scope of this book.

Constraints on the completion of a project may include the technical capability of the company to deliver the project. Cases have been seen of shipyards taking on projects with onerous technical specifications, where the shipyard did not have the expertise or could not find this from a sub-contractor. Ambition and optimism may be supported above realistic assessment of the true scope of a project.

Risks will exist for any project which is going to deliver something new, whether a change in technology, an innovative building, a road or rail scheme, or in the case of the subject of this book, a ship. Inevitably in these circumstances there will be risks. That is, there is a danger that something may occur which prevents the objectives of the project being realised.

It is therefore really important to ask the basic question, what can possibly go wrong? Many projects do not fully explore potential risks to delivery, or completion of the scope, or meeting the budget. All of this can be very costly. Where risks are found, it is essential to determine how they can be eliminated, transferred to someone else or managed. Assumptions may be made about the project, perhaps because of a strong wish to see it go ahead but any assumption is a possible risk. There may be aspects which are unknown for the project and there is always a risk of over optimism, leading to delivery problems.

A stakeholder is anyone who has an interest in or possible influence over the project. All stakeholders have their own objectives which may or may not coincide with the project goal. They can all make contributions, which may or may not be helpful. They have strengths and weaknesses and in some cases legal rights. Whoever they are and whatever their interest, it is essential to identify and take them into account. To the extent it is safe to do so, it is useful to keep them informed about the project so they remain engaged with it.

Obvious stakeholders are the client or clients for a shipyard, and the shipyard itself is the other main stakeholder in the success of the project. Beyond these are sub contractors, suppliers, services providers and others engaged by the customer or shipyard to assist the project. There may be separate funders for the project, such as banks or governments who have a strong interest.

Apart from more or less direct participants, there are also regulators, classification societies, and others with a safety or environmental interest. Local and national governments may have an interest. Beyond these, there is the general public, and possibly pressure groups who claim to represent environmental or other interests. Some of these may misunderstand the goal or possible impact of the technology to be used and be dangerous to the project.

Where a large company is the client, then other parts of that company may have some reason to oppose the project, perhaps because they have an alternative they wish to pursue. Large public projects may have a political dimension, with their progress dependent on support from a particular political grouping.

It may not be immediately obvious who the stakeholders are, so it is important to create a full list. Some of them may be supportive of the project, but others may have legitimate or non legitimate objections. Many projects are straightforward, but it is always worth ensuring that potential stakeholder interests have been considered and understood. It may then be a simple task to manage them, but for some projects there will need to be a considerable effort.

A list of stakeholders should be created. This will identify whether they are internal, either to the project team or the company, or external. The external stakeholders are basically everyone else with an interest in the project. Sometimes the real needs of the customer can be lost, because there may be conflicting needs between the stakeholders leading to "wish lists" of attractive but unnecessary project features. The customer may in fact not be entirely clear what the real needs are, especially if the customer is actually several different interests. A government project may have an intended operator looking for many features, a sponsoring department and a financial department looking for minimum cost.

Typically, internal stakeholders will start with a project sponsor, usually a senior company official who will authorise and support the project. The project manager will manage the day to day operations for the company, reporting to the management and the client. A project team, drawn from those departments which are supporting the project with personnel or technical assistance will carry out the work. Often for ship projects there will be team members from major sub-contractors. Other departments of the company will carry out routine work in support of the project.

External stakeholders will be most importantly the client for whom the project is undertaken. Others may possibly include funding sources, for example direct investors, banks or governments. There are also subcontractors and suppliers to the contractor who are not part of the project team.

Further, and perhaps less obvious stakeholders are classification societies and other marine regulators, for example the US Coast Guard or other national bodies. Governments may have environmental regulators and health and safety regulators. The public, often with pressure groups as proxies, especially for environmental or safety issues. As part of policy, a government, national, state or local, can influence the project.

It is important at the outset of a project to consider all the possible stakeholders and make some assessment of them. This will include, their potential influence on the project, whether they are supportive or opposed to the project. Also it matters to decide how they may be helpful or obstructive in managing the work.

In many cases, with straightforward projects, the review of possible stakeholders will eliminate most of the possible ones. In other cases there may be a need for some political lobbying, or positive publicity to ensure the project does not suffer. Increasingly, environmental issues can be a source of considerable problems, sometimes justified by how the project will affect a locality and sometimes without real justification.

The selection and working of the project team is important, starting with a project manager. Project management is a dynamic process and as described above it should be based on clear objectives and leading to a change. It is multi-disciplinary, is certainly performance oriented and control oriented. The project manager has a temporary role as, although he may be a professional project manager, each project is different.

The project manager's role is diverse and requires considerable management skills and experience. The manager has to be flexible, show initiative, be good at communication, and have general rather than very specific technical knowledge. A project manager is also decisive, and this suggests someone who is outstanding. The skills are often more aspiration than actual, but they can be learned.

Ideally selected for experience, skill and commitment, the project team is often a compromise. It is essential to review the team and seek to fill serious gaps by external resources, training and close control. Project teams are multi-disciplinary and should follow carefully some important selection criteria. These include technical expertise and experience, relevant authority and limited or preferably no other commitments. Team members should be team players, have commitment and not have their own agenda. According to the literature, simple availability of someone should not necessarily equate to selection. In reality it is often the case that the team is made of whoever is available, requiring the project manager to also develop the team.

Risk management is an important topic and is here discussed in more detail. The word risk derives from the early Italian risicare, which means to dare. Now risk is defined as something which creates the possibility of loss and risk management is an important basic input to decision making when a project is set up.

There are numerous risks to a large project, of which constructing a ship is a good example. A risk is any event that, if it occurs could prevent the program or project realising the expectations of the stakeholders as stated in the agreed business case or project definition documents. Examples for a ship project could be late delivery of a major equipment item which then delays the completion of the ship, or failure of a major sub-contractor. A risk that actually happens is treated as an issue.

In the management of risk, there some factors which are critical to successful outcomes. First there are people who will implement the risk management process, who must be trained and preferably experienced in the project management process and in what risks are likely to occur. It is the people who will identify the risks and if this is not done effectively then the rest of the risk management process will be of little value.

The process of risk management is designed to transform uncertainty, which is an input to the process, into acceptable risk as an output. The process must operate within a culture that supports risk management. Concern about risk can be interpreted as pessimism, and many people engaged in projects do have an over-optimistic view of project management.

Two definitions can be made here, first that a pessimist is simply an optimist with better information and second that a pessimist is what an optimist calls a realist. Proposing effective risk management is not taking a pessimistic view, it is simply trying to anticipate any problems that might occur, evaluating them and then taking action, if any is actually required, to avoid them. Identifying risk is a collective activity, to be carried out by the project team, with no suggestion of a possible risk being dismissed until it has been carefully considered. The team should also use others outside the team where a more qualified opinion can be found. It is possible that the identification of risk will show very few really exist, but that is unusual for a large, complex project. It is better to be sure than to miss a possible problem.

Once risks have been identified, then there is implementation of the risk management process, which is the plan and methodology used to perform risk management. This begins by sorting risks into different categories. Three can generally be identified which are transfer risks, avoidance risks and residual risks.

Transfer risks are those that are absolutely unacceptable to the project and cannot be managed by the project team itself. As such they can be only be resolved by a third party for example a qualified sub-contractor. The strategy for these risks then is to pass them to such a third party to take action. As an example, if a ship specification requires very low sound and vibration so that scientific equipment can be operate successfully on a research vessel, then the required expertise will not normally be available to the shipyard design team. So in such cases a third party company with the required expertise will be used by the shipyard.

There is then a new risk inherent in the use of sub-contractors, which is that they may also not be able to deliver the required results, or cannot do so within a timescale and cost set by the overall project.

Avoidance risks are those that must be avoided immediately with some appropriate actions, for example revising the scope of the project or amending the plans for its completion. The company carrying out the project will need to develop a risk mitigation plan and then take appropriate actions immediately to avoid the risk occurring. This presupposes that the design or scope of the project can be amended, although this is not always the case. Collaboration with the customer to identify and develop acceptable alternatives is ideally the initial action to be taken.

Residual risks are all the other risks that can be managed and controlled within the programme or project. It may be decided that some are still unacceptable, in which case the project team should revisit them to see whether transfer or avoidance is the best strategy.

It is necessary to assess both the probability of occurrence and the potential effects if the risk does occur. Then the risks can be ranked as being, high risk, where the

Fig. 4.2 Risk matrix

Probability	Effect on the Project		
	Low	Medium	High
0.8 - 1.0			
0.6 - 0.8			
0.4 - 0.6			
0.2 - 0.4			
0.0 - 0.2			

probability of occurrence and the consequence would be severe, medium risk, with less potential harm to the project and low risk, where the outcomes if a risk turns into an issue are manageable.

One approach to ranking risks is to create a matrix, the axes of which are the probability of occurrence and the severity of potential outcome. This makes assessment easier and also provides a useful visual interpretation of the risk situation. Figure 4.2 illustrates a risk matrix.

Once the risks have been identified and classified, the remaining residual risks should be recorded in a risk log. This lists each risk, its potential for harm and probability of occurrence. Depending on the severity of each risk, considerable data may be collected and recorded to assist the project management team. The log will also assign an "owner" to each risk, who will be responsible for managing the risk, recommending and taking agreed actions and reporting to the project management, especially when an occurrence seems likely. From the log, each risk will then have a risk management form developed, with actions to take if the risk shows signs of occurring.

As the project starts and progresses, and as and when risks emerge as potential issues, the risk owners will monitor the necessary actions as they are taken. It is essential to maintain the risk log through the life of the project. It is not simply completed at the start and then left. Then as the project unfolds, the team will update the log with changes to risks and also with any new risks that emerge. These go through the same assessment process as described above.

Even risks that have been dealt with remain on the register, as the register will become a useful tool for future risk assessments on new projects. It is also sometimes the case that risks re-emerge having apparently been removed earlier.

Some form of risk management exists in companies whether it is formal or not. In the worst case it may be reacting to an issue that has arisen unexpectedly. This risk management is generally based on the past experience of the project manager and team, so may be incomplete if the formal process is not followed. Without a formal procedure, experience may not be shared and a formal procedure importantly gives an opportunity to manage future project risks more effectively.

There are some very sophisticated risk management models, which can predict likely outcomes, compare risks and quantify the outcomes. However they all depend on the identification of the risks, and their consequences in the first place, and then on the assessment of probability of occurrence, If any of this is not carried out correctly then the models cannot offer a realistic answer. Simpler methods with sound data may give a better result.

Although this was initially said about foreign policy it works for project management. In any project there are "known knowns" where a requirement is clear as is the means of meeting it. This is the ideal situation where the project is fully defined and within the capability of the project team. "Unknown knowns" where a requirement is not clear, or something unexpected happens during the project, but where a solution is easy. "Known unknowns", are any residual risks which have been identified and where a solution has yet to be found. "Unknown unknowns" occur when a problem emerges entirely unexpectedly, and presents a serious challenge to the project team.

If the project can be fully defined, the capabilities to carry it out are confirmed and any risks have been eliminated or at a minimum reduced to an acceptable level, then a decision to proceed can be made. The second project phase is under the overall scope of planning. This includes the determination of when the individual elements of the project will be carried out, including the design of the product, procurement of materials and other external supplies. It also includes the acquisition of the workforce, the equipment and the design of the processes required. Each of these is discussed in later chapters.

Part of the planning requires a work breakdown. This is critical to project success, because it dictates how the work will be organised and planned, and also how progress is measured. There is an initial breakdown into main elements, then a further breakdown into sub-elements. Ultimately the work breakdown identifies small, short duration work packages, which make the project manageable. Once the project is not only defined fully, but also all the requirements in planning and resources have been put in place or as a minimum identified and are known to be available, then work can start.

The third phase is project execution, which if all the preceding activities have been completed successfully should be straightforward. In reality the project manager and his team are there specifically to manage situations where the project begins to go off course. Progress monitoring to ensure the project runs according to plan is discussed in Chap. 17.

When the work is nearing completion, preparation begins for the closure, the final phase of the project. A key issue is to avoid project creep which is any tendency for the customer, or even sometimes the project team, to try to make changes to the project scope. To avoid this it is necessary to set up an acceptance procedure and criteria for completion of the work. A formal closure meeting, in some agreed form, is needed where agreement between all relevant stakeholders is reached.

After completion the project team should make a formal, in-house evaluation of the project. This is to identify good practice from the project, to avoid future mistakes by careful analysis of anything that went wrong and to provide a learning experience for the whole team.

Evaluation includes active evaluation during the project, because it is often possible to correct mistakes and adjust processes as the project unrolls. Evaluation includes a review of the project management process and also whether the project was a technical success, although keep in mind that a technically successful project can still bankrupt the company which carried it out.

However the project is portrayed to the outside world, to the rest of the company or any other interested parties, the team should be sure that they to know the "Truth". This need not be shared beyond the team, especially if that may be dangerous, but should be acknowledged by those who participated. Knowing what really happened is an important part of the learning process for the future.

Chapter 5
Shipbuilding Company Strategy

However beautiful the strategy, you should look sometimes at the results
—W. S. Churchill

Using as a basis the activity map described earlier in Chap. 2 of the book, the development of the company strategy for a shipyard can be outlined. This is the first stage identified in the activity map as a necessary precursor to a ship project. Essentially the strategy should be in place prior to any ship construction, as it provides a context for the construction process. The strategy has the characteristics of a project and can effectively be managed as such, or as a programme of several projects, including for example facilities development.

The strategy should begin with an overall corporate objective or goal. Such a goal is sometimes expressed as a "Mission Statement". There can be a wide variety of mission statements, from relatively unspecific to highly focussed. Statements such as "to be world class" or even "to delight our customers" are not really helpful as a starting point for developing how the company will be managed. Others can be very specific, such as to reach a specific market share, and this is preferable. A low key but useful approach to the corporate goal is something like;

"To still be in profitable business in five years time".

A small number of very large shipyards may be seeking a large market share as their goal. This reflects the tendency for a small number of shipyards, often in one country, to dominate the shipbuilding market for a period of time. In other cases a government supported industry may be seeking a specific growth target and this is also a useful goal. For the majority of shipyards, operating in a market which is always difficult, the goal of continuing successfully is realistic.

The development of the strategy can be regarded as a project, because the various attributes of a project are present. There is a goal which can be, indeed should be, quantified and expressed in specific outcomes. These include a level of income, and ideally a profit. There is a timescale, typically five years although it can vary considerably, depending on the market in which the shipyard will operate. For military ships the timescale will usually be much longer, reflecting the lead time for new ships. For

G. Bruce, *Shipbuilding Management*,
https://doi.org/10.1007/978-981-15-8975-1_5

small ships, the time may be shorter. In any case there should be a point in time when specific objectives are intended to be achieved.

The strategy development is unique for any time period as the external conditions in particular will be different for later time periods. The strategy is specifically intended to benefit the company, which for the purposes of looking at a project to develop the strategy is its own customer for that project.

There is some routine work in the project, but the set of activities carried out make the strategy development non-routine. There is a complexity due to the many interacting factors in the strategy. There are definitely risks in the development and a particular example of risk is in the making of forecasts about the future market. There will be sub-projects in the overall project, for example the development of new products or the creation of new ship construction facilities.

The development of the strategy will require inputs from all the main functional units as shown on the activity map. So starting with a corporate mission, a set of linked strategies is developed for each shipyard function or department.

A good starting point is the definition of a planned product mix, which is the set of ships which the company management intend should be built. This requires some market analysis, to look for ships for which orders can be received at prices which are acceptable. This sets a goal for commercial activities, beginning with the market.

An analysis of the product mix is then required to identify the various main interim products, the items which make up the ship. Of these interim products, security items are those which the company strategy determines are essential to be produced in house to minimise project risk. Often the hull of the ship is seen as the most important of these items, because this is the most significant feature of the ship. Alongside the hull will usually be the piping systems and some of the main engineering elements.

Not all shipyards take this view and in a few cases the hull is actually seen as a low value commodity which can be built almost anywhere and so the lowest cost acceptable hull from a sub-contractor is seen as the ideal solution. This brings up the question of what other elements of the overall ship production process should a shipyard carry out. Sub-contracting is increasingly popular, primarily because it is seen as a cost-saving opportunity. Decisions are made on which of the items in the ships to be built should be made by the shipyard itself.

High value items are worth producing in-house because they earn income and create profit. For some specialised ships, the engineering equipment is of high value and a shipyard may concentrate on this rather than the actual process of building the ship hull. Dredgers and other ships with an expensive or sophisticated outfit come into this category.

Most shipyards will sub-contract items such as cleaning, painting and scaffolding because these are low value items which earn little or no income and are certainly not usually profitable to the shipyard. Many shipyards also sub-contract outfitting, especially where the requirements fluctuate through the life of a contract, which makes maintaining the capability full time uneconomic.

Strategic components are those which the company decides it must make in house in addition to the security items mentioned earlier. All other components are potentially items to be sub contracted, so make or buy decisions are needed for these other items, depending on availability of suppliers, costs, risks and other factors. Key facilities are then designed for the production of strategic components. All the shipyard facilities are planned for the in house production and suppliers are identified for bought in items.

The decisions which are made should be reviewed from time to time in case the cost structures have changed or the risks have altered. Sometimes suppliers may go out of business, or become sufficiently busy to be unable to meet a shipyard's needs.

Transaction costs are an issue where a shipyard chooses to sub contract work. These are the actual and hidden costs of using sub-contractors, for example contract costs, which can apply to any supplier to a shipyard. If work is carried out in the shipyard then there is a need simply for a management instruction to do so. However if a sub contractor is used, then there must be a written contract. This may be simple for a regular sub contract, but for a new supplier or a complex piece of work there can be a significant cost in arranging for the work to be carried out.

There may also be a need for increased insurance, to cover the transport of the finished piece of work. There may also be a risk of failure to deliver, which can carry a large consequential loss if a ship project is delayed as a result.

Quality assurance is another potential sub contract problem because of costs for periodic progress inspections at the contractor's site and inspection on delivery There is always the danger that the work may not be satisfactory. Careful selection of the sub contractor can mitigate these issues, but there is immediately another cost in identifying and choosing who will carry out the work.

The decisions on what ships to build and what strategies to adopt for the production or procurement of items for them will inform the rest of the strategy development.

Considering first the strategy for the production function, which carries the largest cost because of the numbers employed and facilities requirements, this requires documentation on standards, quality and productivity targets to give a complete definition. The production strategy, as with the other functions, must be reviewed over time. This is because of changes in products, changes in the technologies available, changes in the competitors of the shipyard and changes in the environment in which the company operates. The strategy can be expressed as how the company operates now, how the company will operate in the future and importantly, how the change in operations will be managed.

There is a danger in using historical data for future planning unless it is based on a competitive performance. The performance of the best shipyards keeps improving, so to remain competitive a shipyard must plan for future improvement. Past data is also dangerous as a basis for planning if it is not based on stable production processes, for example if the products of the shipyard keep changing, so it is based on high product variety.

Another potential problem with past data occurs if the workforce turnover is high so that knowledge is lost. If the performance is based on a poor work breakdown structure, inadequate planning or any other deficiency in the operation of the shipyard activities, then the data available is again suspect.

As a result it is essential to design the production processes at all levels. There is a need for a culture of continuous improvement, so that the people employed are searching for ways to improve their performance. This may coincide with making the work easier. It should start at the bottom, so the workers in any work station or group, who should know more about their work than anyone else, are encouraged and supported in seeking improvements.

For the management there must be an awareness of competitive performance, which requires monitoring of the shipbuilding market and which shipyards are winning orders.

There must be a product mix which is appropriate to the company, that is to say the ships to be built are suitable for the dock, for the cranes which are available and the capabilities of the shipyard in general. In the early days of VLCCs, a number of shipyards attempted to build these vessels, much larger than previous tankers, without a significant change in facilities. Although they could build the ships, slowly and with difficulty, they could not do so for any possible profit. Other shipyards have built ships in docks which were too small, or have joined half ships afloat. While overcoming the technical challenges has often shown ingenuity, the results have often not been profitable.

There is also a need for close fit between the product designs and the production processes. That is, the equipment and processes should be designed for the ships. Again examples can be found of shipyards with, for example a panel line for flat, stiffened steel panels choosing to build bulk carriers with corrugated bulkheads, rendering the expensive panel lines useless for the programme. A balanced set of facilities is required, so that effort and investment is spent in those activities of greatest importance and where man-hour and other savings can be made. A quality assurance programme is also essential.

For the product to be made at an acceptable cost and within a sensible timescale, there must be a rational product work breakdown structure, creating a necessary match between facilities and product. Once a building programme is under way, there must be good monitoring of progress and follow up action to correct any problems which are discovered. Finally, the results of the building programme must be assessed to offer feedback to the future strategy.

Given a goal for the strategy development project, then each function in the shipyard will have a contribution to make in its achievement. It is useful to start with the commercial aspects. Any marine production facility can only operate in the context of the market demand for its products. A large shipyard may carry out its own research into the market, or may commission a specialist research company to work on its behalf. Based on all the above, the outcome from the commercial function will be a forecast for the demand for ships which the shipyard may undertake.

From the information gained, the technical design function can prepare outline designs for potential products. These may be new products, incorporating radical developments or may be current products with small updates. The designs will then be the basis for marketing the company to potential customers and also a basis for a review of the facilities available. There may be a need for changes, new facilities, re-training or new equipment.

There will occasionally be the development of a new shipyard though this is comparatively rare since the turn of the twentieth century. Because of the long life of the main facilities in a shipyard, there are many shipyards in the world which were developed since around 1960. Major surges in development have taken place, in the early 1970s as a result of high demand for large oil tankers, and in the late 1990s as a result of predicted demand increases. Nations new to shipbuilding have emerged, and as a result developed large ship construction capacity.

Many other nations have re-developed their shipbuilding industries to participate in new markets. These have mainly been on a smaller scale of development, although a major exception is in Europe, where a number of existing shipyards decided to concentrate on the construction of large passenger ships.

Smaller changes to shipyards happen on a more or less continuous basis, making incremental changes to facilities and technology. These are as a result of the market analysis, the ageing of some equipment and availability of new production processes. The main facilities found in the industry, particularly large dry docks for construction, large cranes and the buildings associated with assembly operations have long lives if properly maintained. For the purposes of describing facilities development, the whole process will be outlined, with the actions needed for a new shipyard. A subset of these will apply to smaller developments.

If the development of a shipyard is considered as a new project then the main factors which play a part in determining the layout and how the shipyard will operate can be identified. First, the product mix is developed from market research and then the proposed build strategies for the ship products in the mix. The strategies will include the breakdown of the ships into interim products, in particular the hull block breakdown. This will determine the speed of construction, which is dependent on the market expectations, volume of production and the cost and income balance expected. This breakdown will also determine the main crane capacities for assembly and ship construction and the other major material handling requirements.

Given the blocks selected, there will then be an interim product hierarchy, identifying smaller assemblies which will be a basis for the design of individual work stations. The numbers of interim products to be made will determine the size of the buildings, main equipment requirements and transport needs. The analysis will finish with the sizes of raw materials and other items from suppliers. The size of steel plates and profiles is of particular importance, because larger steel material allows larger assemblies and blocks to be made with minimum joining and welding of parts.

Supplier capabilities and locations also need to be considered to ensure reliable supplies. This is important for the outfitting of the ships and will also affect the decisions to make or buy ship interim products and equipment. It will also determine the scale of outfitting facilities which are required.

The proposed site for the shipyard is of major importance. Often there are several potential sites available, either existing shipyards with space to expand, "brown field" sites where the area has been use for industry previously or "green field" sites which will be developed for the first time. A new site should ideally be close to communications, a potential labour force and possibly suppliers. Alternatively the provision of some or all of these can be part of the total development package.

The physical characteristics of the site are important, including ground conditions particularly for the support of large loads and the construction of major facilities especially dry docks. The location is important largely from the point of view of the infrastructure requirements. Infrastructure includes access, labour availability, other supporting industry and communications.

The level of technology to be adopted is important for a cost-effective facility. A low level of technology may be appropriate where there is to be a low level of throughput especially in a low wage economy. A high level of technology is appropriate for high value products such as large passenger ships. It is also appropriate for high volumes of production and where shipyards are located in high wage economies.

The critical part of a marine production facility is the final construction area. The ability to progress the required series of products through final construction largely determines performance of a facility in terms of the number of ships which can be built and therefore the overall income which will be available.

The choice of facility is wide. It is usual to find a dry dock, which allows for a safe and controlled float out of a finished ship, but there are alternatives. A sloping slipway is the traditional facility, and is generally less expensive than a dry dock. There are some operational disadvantages from the necessary dynamic launch, and it is more complex in a non-tidal region. A flat construction area for load-out of the completed ship onto a shiplift, or occasionally onto a floating dock is sometimes adopted. Any of the choices may be open to the elements or may be partly or completely covered for weather protection. Cover is usually associated with high value ships, often military or passenger vessels, or for smaller ships where there is a poor climate, for example significant snowfall.

The final construction site is the area of highest capital cost and therefore important. The handling and storage of materials, assemblies and blocks also has an effect on efficiency. Shipyard sites are generally large and spread out so movement of items and people around the site can be a large cost factor. The method used to move final assemblies to the ship construction area is a major cost element. Alternatives include level luffing jib cranes which requires the assembly location for the assemblies to be close to the construction site. Then there is a direct transfer using the construction cranes. The same is the case for portal, or goliath cranes used for construction. Where the construction of the ship takes place under cover the overhead cranes in the buildings can make a direct transfer to the ship under construction.

To separate the assembly and construction areas is a solution which offers much greater flexibility in operation, and so most shipyards have adopted self-elevating transporters. These vehicles are hydraulic and can move, elevate and lower a load deck and manoeuvre, to take large loads from assembly to interim storage and then deliver the assemblies direct to the dockside cranes. Some other ground level systems are used to move very large loads, either wheeled systems or sliding or "walking" systems.

There are a number of specialised facilities which may be sub-contracted and the manufacturing strategy for a facility will determine whether or not these are on site. This will depend on various factors, including the availability locally of suitable sub contractors, the required volume of production, degree of specialisation and the relative cost of production in house. Typical sub-contracted facilities are electronics, HVAC, painting and most equipment installation.

For interim products with a high throughput, a dedicated facility may be required or desirable for their production. Many workshops and work areas are general purpose. A typical example of a dedicated facility for high volumes of production is a flat steel panel production line for large ship hull blocks. Although dedicated facilities often represent opportunities for high technology, they can result in what are called “islands of automation”. That is a small facility which does not impact much on the overall production efficiency but which requires investment and a lot of management and engineering attention which could be better used on more man-hour intensive areas.

The production technology depends on how the elements of production are used. These elements are people, materials and machines. Depending on the process and the stage of a project, various combinations can be used. Any of these elements, or any combination of two, can be fixed in position or can move. For example on the ship a work group installing equipment will move to the location, and take the equipment and tools they need with them. In early production, materials may be moved to a fixed work station with the workers and equipment in place.

The facility layout process determines the disposition of the work areas on the site. The criteria for a good layout are usually expressed as a list of good attributes. Some of these may not be appropriate in specific cases, but they are sound starting point for the layout of a shipyard and the facilities within it.

The first is to have minimum site area, consistent with space for the facilities needed, storage for items between facilities and space for possible expansion at a later date. Land is usually expensive, especially adjacent to water, and earth moving or reclamation and other preparation needed also carry a significant cost.

Alongside this is to make maximum use of the area and volume of the buildings which are available. Except for possible future expansion, the land area should all be in use for a specific purpose, again with the objective of keeping costs as low as possible. Building heights should be kept as low as possible, subject to the operation of overhead cranes and safe lifting of loads. In particular for storage areas, the trade off between building height and area should be considered, so as to obtain the most capacity again for a low cost. High racks for storage is one means of maximising use of available volume.

A layout with unidirectional flow is desirable. This is so that materials, parts, assemblies and units move progressively through the various processes. How this is arranged will depend in part on the site configuration, unless a new, rectangular green field site can be made available. The progress does not have to be in a straight line indeed with the objective of minimum site area it is better to arrange a flow around three sides of a square. This can be used to provide storage in the centre of the site with flow round it. Figure 5.1 illustrates a basic layout for a shipyard.

Such an arrangement as above also fulfils another desirable attribute, which is to have minimum materials handling. That is to keep the distance travelled by items as they progress from process to process as short as possible, and to keep the movement of transport to a minimum as well. The need for minimum movement, including for the workforce, and the contrasting need for large areas and space for potential expansion, make the layout development process challenging.

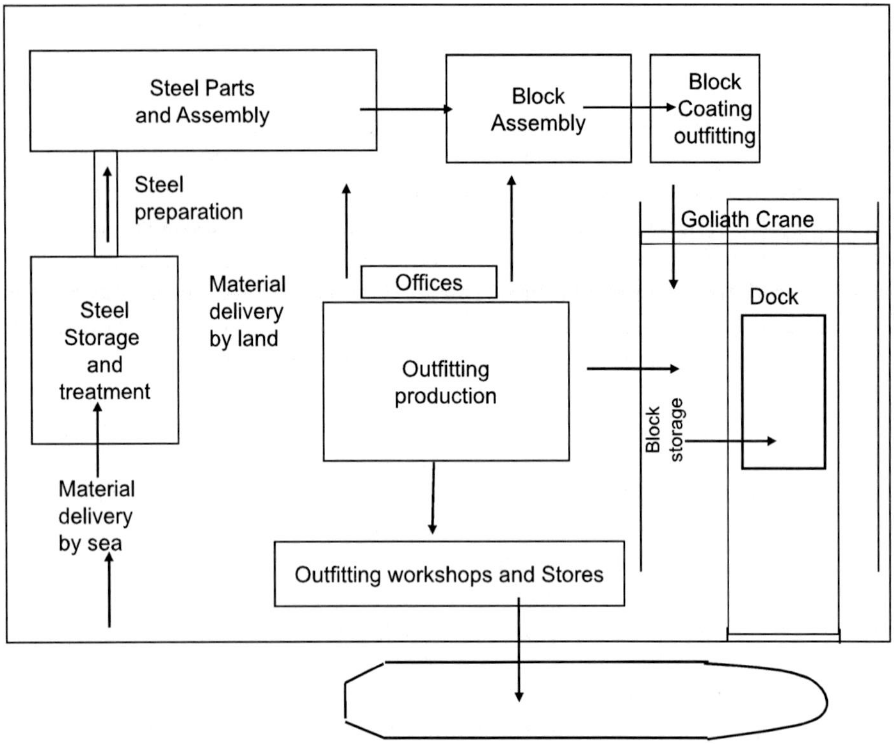

Fig. 5.1 Typical shipyard layout

Flexibility in operation is desirable. Although the shipyard will be developed with a specific product, or a small range of similar products in mind, consideration should be given to how the market and consequently the product range may develop. The ability to switch to an alternative product with minimum disruption is a valuable characteristic for a shipyard.

The attributes mentioned so far will help to provide easier supervision of labour and processes. Visibility of progress will be good, and the distance to any part of the site will be short, encouraging direct supervision both locally, and by management.

In order to quantify the advantages of a particular layout, a systematic method should be adopted for its development. The basic procedure for developing layout is described here.

The first and fairly obvious stage is to identify the requirements to be satisfied by the layout. These are the required volume of production for each of the main facilities and then for the more local processes. This requires an analysis of the ships to be built, creating a build strategy with a breakdown of the whole ship into interim products. The numbers of these are then determined.

The work content of each product can be estimated, which combined with the volume will allow the number of workstations to be decided in order to produce the required volume. The size of each workstation will be decided, with work areas, service areas and access, and also the storage requirements for interim products entering and leaving the work station. Several workstations will usually be located in one building or work area, which will allow building sizes to be determined. This information is a large part of what is needed when considering the overall site requirements.

The proposed flow of work through the work stations will indicate the relationship between the stations. Those where the interim products are processes in a sequence to create a finished assembly will need to be located close together, possibly in the same building.

There will also be inspection requirements during the processes and of the interim products, mainly assemblies, before they proceed to the next stage of work. Area is required for this purpose and also may be needed to provide a repair location, for example if a process is known to be sometimes unreliable. Of course if this is the case then production engineering should be looking urgently for means of improvement.

There are health and safety considerations to be considered in the layout. For example some storage must be at a distance from workers using hot processes, such as flammable gasses. Some processes should not be close together for safety reasons.

Access and a holding area is required for the material delivery to site. Parking for vehicles may be needed, depending on the unloading arrangements. Also inspection of items may be required before they are accepted into storage in the shipyard.

Waste disposal is an important consideration, increasingly so as concern for the environment grows. Space is required to collect waste from the shipyard processes, ideally immediately after the material concerned is used, for example scrap steel after cutting or packaging after equipment is used. Waste sorting and storage prior to removal from site is also required and provision is needed for this.

Where sub-contracting is in use, some facilities will be needed for their operations. These may be substantial buildings, perhaps even with some equipment. Storage for the subcontractors' materials may also be needed to facilitate the efficiency of their operations.

Apart from production facilities, which are the main interest here, provision is also needed for offices for technical staff, managers, supervisors and non-production staff. Some of these may be best located close to relevant work stations and in the buildings, but others will need to be at a distance from the work areas. Other provision may be made for transport of staff and workers to the site, around the site, for public transport, shipyard provided transport and parking for private vehicles.

There may be canteens, changing rooms, washrooms, even social facilities, depending on the national and local policies which prevail. First aid provision, which may be a small ambulance room with first aid trained staff or a more complete medical facility is usually found. There may be other requirements for specific shipyards, for example in a remote location living accommodation may be needed.

As with production facilities, all the additional buildings and areas which are required will need to have estimates of their space requirements. A list can then be developed of all the facilities, their areas, lifting requirements, building dimensions and other characteristics.

It is then necessary to identify the relationships between each pair of facilities on the site. This is a systematic look at each pair, even when it is likely that no specific relationship will be identified. The relationship is dependent on whether there is a need for the facilities to be close together, or whether they can be distant. In some cases the facilities will definitely have to be well apart.

In addition to materials flow, facility design also takes into consideration the movements of people, services and information.

The closeness of relationships between the various facilities can be quantified, usually on a simple scale of 1–5, where 5 indicates a need for facilities to be close and 1 that there proximity is not important. There may also be a value of −1 to indicate necessary apartness. Those facilities which need to be close are located first on the site plan, taking account of the site shape. The position of major facilities like a dry dock will usually be predetermined by the obvious need for proximity to the waterfront, but also the ground conditions. Generally the major facilities will be located first, based on the site conditions, access to the site and a general preferred flow.

The facilities which are required to be close to others are located first. The closeness can be as simple as the number of facilities which need to be close to the first one under consideration. The total sum of the one to five closeness scores when their relationships are considered will dictate the first facility. Then the less critical facilities in terms of closeness can be located according to their relationships.

The layout can also be optimised against an objective, such as minimum travel for major loads or minimum total tonne kilometres of transport. Other considerations include the need for large cranes which can dictate closeness of facilities to save duplicate costs. Facilities using a single expensive, heavy crane may be close. Worker access may be another important criterion.

A relationship diagram is a common approach to developing the layout, using an objective or subjective measure, to determine the relationship between the facilities.

Where the production volumes and facility sizes are known, the relationships can be quantified by calculating the number of journeys to be made or the total quantities to be moved. For this the proposed means of transport must be decided. The information can then be presented in the form of a network, which is the basis of a layout definition. The network nodes represent the facilities and the linking arrows the number of journeys.

Figure 5.2 shows a matrix with the facilities A to G. The numbers represent the closeness score for each pair of facilities, which may be the total of closeness scores of 1–5, or a measure of the movements between them. Once the matrix has been constructed, the network can be developed. Those facilities required to be close can then be moved until a suitable concept layout is developed. This is later used as the basis for the location of the facilities on a site plan.

There are a number of basic shipyard layouts, although the precise arrangement will depend at least in part on the available site. These layouts are based on the main construction area which may be a berth, a dock a flat area or a shiplift. They are also based on the historical development of facilities which in many cases have changed to match increasing ship size, to meet shorter delivery times and to reduce

	A	B	C	D	E	F	G
A	-	100	70	25	-	-	-
B		-	90	150	75	45	-
C			-	50	11	68	-
D				-	78	55	-
E					-	60	22
F						-	36
G							-

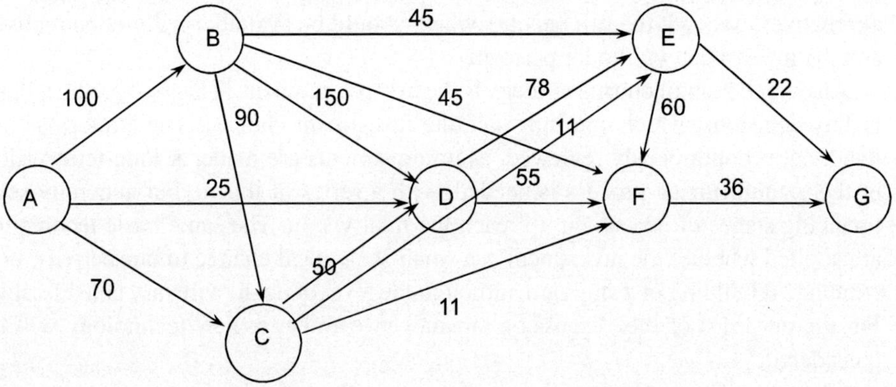

Fig. 5.2 Shipyard movement analysis

construction costs. Other shipyards have barely changed in their basic arrangement for many decades, if they concentrate on small and medium sized ships.

Around 1950, shipyards were effectively the direct descendants of late nineteenth century and early twentieth century shipyards. These in turn were descended from the yards used for wooden shipbuilding in the nineteenth century and earlier. The shipyards were typically on a river bank and with many berths where it was desired to achieve high production levels. There was piece by piece construction of the ships with the most of the outfitting started after the steel hull was complete and launched.

During the late 1950s and the 1960s, many shipyards developed from the older shipyards, with fewer building berths, with larger cranes. Ships were built using unit construction and much of the outfitting was still done afloat. There were always exceptions to the usual commercial shipbuilding, with large passenger ships and military ships sometimes using dry docks and some larger cranes.

As the size of ships continued to grow in the 1960s and into the early 1970s, and ship sizes, especially for oil tankers increased, developments were made on greenfield sites. Such shipyards were generally designed specifically for the larger ships and were equipped with large cranes.

Also in the 1970s a number of shipyards adopted undercover construction with all or most of the activities located in a single building. This provided a compact layout with easy transfer of items from process to process. Moving entirely undercover also gave weather protection, and allowed uninterrupted work in wet climates.

The choice of layout for an existing shipyard is often governed by the existing facilities and the development must be arranged to make the best use of these. For a green field site there is usually more freedom, subject to the site limitations.

It is always necessary to make a serious appraisal of a proposed investment, not only to avoid a low return on the money spent, but also potentially enhance a sound investment. If ad hoc decisions are taken, they can lead to two major problems. First, investments may be incompatible with later investments, so they should be part of a longer term strategy. Also, an investment may simply be less effective than alternatives. Second, the full benefits which should be available will not be realised and the investment will under perform.

An overall manufacturing strategy is the first pre-requisite. This will include make or buy decisions, which can influence the investment choices. The strategy is not fixed, but is continuously reviewed, as improvements are made. A long-term vision of the manufacturing process is needed, with a series of linked, but generally self-financing stages of investment to reach the final vision. The same basic techniques are applied whether the investment is a small scale local change to one activity, or a complete rebuilding of a shipyard, although the level of detail will vary considerably. For the purposes of this discussion, a small investment in new technology will be considered.

The cost savings which will come from the adoption of introducing some more advanced technology come in a number of ways. Improved local productivity is the first, so for example a better welding system will expect to produce more in a given time, lowering the cost of each product. Fewer personnel may be needed for the same output, and there may be reduced scrap and re-work. Other potential savings include lower stocks of materials or lower work in progress offering a smaller working and storage space. Shorter lead times, improved adherence to schedule, improved control, better continuity of production and integration of departments may also be benefits.

Other benefits may be regarded as "unquantifiable", but in practice if the production system is adequately modelled, they can generally have a value placed upon them. For example, the ability to produce with higher quality, along with an image of a technological company may be a factor which assists sales.

A general model for investment is used. First, all of the costs and benefits from the investment must be identified and listed. The lists below are not necessarily comprehensive and not all these potential costs will apply in all cases.

Investment costs	On going operating costs which may be higher or lower than currently found
Feasibility study	Labour costs
Capital investment	Material costs
Project team costs	Inventory
Consultancy	Maintenance
Training	Insurance
Commissioning	Power
Lost production (temporarily)	
Redundancy costs	

Benefits are typically higher productivity, lower labour costs, reduced building costs and reduced storage space requirements.

Several ways to make the investment appraisal are available. The first is simple payback which is the time taken for the savings from the investment to equal the initial cost. Consider two alternative investments, with different costs and benefits, which need to be compared to decide the best. The cash flow for the investment cost is calculated, showing the initial investment and other on-going project costs. The cash flow from all the on-going savings is also calculated. The final summary figures are tabulated typically for up to ten years while the investment is operational, to show all the costs and benefits which can be identified.

For an initial investment of $100,000 with annual savings calculated to be $25,000 the payback would be four years. After ten years the cumulative saving would be $150,000. This may be acceptable or it may need to be compared to alternatives. A second investment of $60,000 with annual savings of $20,000 would pay back in three years which is shorter, but after ten years the total saving would be $140,000.

The smaller investment has the shorter payback, so may be preferred. However over ten years, the other is better. The cash flow can be discounted to take account of the time value of money. The discount rate can be an accounting rate of return, the net present value or an internal rate of return. The net present value (NPV) assumes a discount rate which reflects the financial requirements of the company. NPV is then the current value of the future costs or savings, based on the discount rate which assumes that the value of the money in the future is less than its present value. This percentage is applied to the future cash flow. The NPV value is then the discounted total savings, less the initial investment. Depending on the discount rate selected and the financial return demanded, the final value of the investments can change.

The selection of an investment from a number of choices is often not simple, because the competing investments may have different timescales. They may also have very similar rates of return. The calculation depends on the quality of information and this may vary since the process is attempting to predict what will happen several years into the future. Over optimistic assumptions are often built into investment appraisals. The selection of the method of discounting and the preferred investment return can affect the outcome.

Because the appraisal is looking onto the future, any of the variables in the calculations may change, for example the labour costs may increase, the output may be lower or the accuracy may be less than expected so there is increased scrap and rework. In such cases, the return on investment will change, and a sensitivity analysis is essential to consider the potential scale of the changes. Generally, a company will look for a minimum rate of return for any project and will not make investments unless this threshold is reached.

There are other real world factors which need to be taken into account before an investment decision is finalised. Taxation is important as is potential tax relief on the investment. Regional and other development grants may be available and other government support can sometimes be found for novel technology. Ultimately, any investment must make a sound financial contribution to the future of the company making the decision.

As with any project, some later evaluation is necessary. For a smaller investment, whether productivity increased as a result or costs were lowered can readily and quickly be determined. For a major redevelopment the final outcome is far into the future, unless there is a radical change in the economic environment of the company. Often, because a large sum of money has been spent, there is some reluctance to make an objective assessment of the value of the investment.

Another issue for the management of a shipyard is the decision on the capacity of the operation. The capacity is the ability to carry out a process and is a function of the resources available. That is what level of production capacity should be provided for, in terms of labour force and equipment, to achieve the planned output. Matching the operational capacity of a production system to the demand can be difficult, in circumstances where future demand is uncertain. That is really any time beyond the current order book. It is also a problem where resource requirements are uncertain.

Because the construction of a ship takes a long time, capacity is predictable in the short term but less so into the future. In the long term capacity planning is part of the strategic development, the product mix definition and facilities planning. Some contingency planning is needed and alternative industry scenarios may need to be considered.

In the medium term it is about matching the capacity of the shipyard to the current order book, typically over two years. The total labour force required and the extent to which subcontracting is to be used need to be determined. In the short term, the requirement is the management of the ship construction processes. This includes scheduling, production control and inventory management.

Measurement of capacity is problematic in marine production. Because of the variety of ships, both types and sizes, and variations even between similar ships, a precise measure is not available. Available measures include the number of slipways or docks, the size of the workforce or the past production levels, for example steel tonnage in a year. However none of these is really satisfactory as there are often variations in requirements, which make measurement more difficult. A recognised global measure is compensated gross tonnes (cgt) as discussed in an earlier chapter. While not perfect, the measure does allow a consistent measure of capacity which is useful for comparison purposes.

Managing capacity is complex. It is relatively straightforward to determine average demand levels and these can be used to determine average capacity. It is then necessary to deal with short term variations, since if demand rises, then the organisation is under-resourced and if demand falls, then there is under-utilisation. The cost of either is significant because organisations can not carry surplus capacity and the use of sub-contracting can also be expensive.

Further, if the type of ship to be produced changes, then the balance of the work-force may shift. For example, more steelwork and less outfitting may lead to a shortage in the one activity and a surplus of capacity in the other.

There are options for managing changing capacity requirements for some businesses. These include maintaining excess capacity, refusing some business or increasing prices. Unfortunately these are not always rational or feasible, and for marine industry, the options will be less because for several reasons. First, the industry works on longer timescales than most businesses, so short term adjustments can be difficult. Secondly there are high levels of labour skills, and it is often difficult to lay off then reinstate workers, because they may be able to move elsewhere. It is possible in some unskilled and sub-contracted areas of operation, although it may still be risky. Thirdly and most importantly, there is fierce price competition in shipbuilding almost all the time.

Dealing with capacity problems in marine production does have solutions. In the short term, periods of weeks or months at most, capacity can be increased by the use of labour overtime. It is not usually sensible to use overtime on a systematic basis. In the medium term, looking up to one year ahead, it is possible to sub-contract elements of the work. Sub-contracting on a very short timescale is often not practical.

In the long term it is possible to re-structure the organisation, which may include new investments and significant training for example to be able to construct a different product type efficiently. It may also be feasible to adopt more automation in the production processes.

There are three basic options for any business which is trying to manage changing capacity requirements. The first of these is to maintain excess capacity with the consequential costs of the unused portion of the total capacity.

The second option is to refuse some business, which is not generally an option for a shipbuilder. For a mass producer of small items, the volume of production can often be managed. For a shipbuilder the volume is set by the binary nature, either an order is won, with major resource requirements, or lost leading to no requirements and of course no business. It is not possible to maintain an output stock to satisfy demand fluctuations because the product is a ship which is either completed and sold, or is not ready. Some stock is possible as inventory at the interim product stage so that fluctuations can be managed to some extent, but this is limited by suitable storage facilities. It is not possible either to react to demand by keeping the customer waiting which can be done for example in retail organisations.

Increasing prices, but this is also not a real option for a shipyard, because the price of the ship has been set in a contract, or if the contract is at a negotiating stage then trying to charge above the market price will almost certainly mean the business is lost.

It can be seen that options available to many businesses are not always rational or feasible, and particularly for the shipbuilding industry, the options may in any case be less. Unlike many other businesses, it is not possible to maintain an output stock to satisfy demand fluctuations. The shipyard deals directly with the owner or operator who is the end customer. There is no distribution chain or retail element in the business, with the exception of stock being possible for small craft.

In determining capacity for the future, the starting point is a forecast of future demand, which has all the problems associated with market forecasts. It is possible to forecast overall market demand, but not for specific products in a large, made-to-order business. Product life cycles need to be taken into account, as ships will change in some respects over time. Therefore some contingency planning is needed and alternative scenarios may need to be considered.

There is also competition to manage. Even companies which dominate niche markets may have problems, except possibly for cruise ships where there are very few competitors.

After the review of demand, resources availability must be considered. These may vary in different ways. Machinery and equipment deteriorates over time and labour forces age, and apprentices may not be in place. If a consistent product mix is maintained, losses can to some extent be balanced by learning curve effects and planned performance improvement.

Learning curves are based on empirical evidence originally from aircraft manufacture at satellite facilities in the USA. The concept is that the repetition of a consistent task leads to improvement in performance. Learning curves give a theoretical basis for this improvement. They propose that there is a consistent reduction in the time taken to perform a task, each time the number of tasks completed is doubled. So, for a very simple example, if the improvement rate is 75%, and the first task takes 1.0 h, then the second task will take 0.75 h The fourth task will take 0.75 * 0.75 h, which is 0.56 h, the eighth will take 0.42 h, and so on.

A typical learning curve is shown in Fig. 5.3.

Although the original theory was based on large numbers of products, rather than a relatively few ships, the learning curve can be identified for many shipyard processes, and for series of identical or very similar ships.

Any production system which is designed for a special purpose will have an economic operating level. That is, the system will be designed for a given capacity and if there are larger or smaller volumes of production, the result is an increase in unit cost. If the production is lower, then the same fixed operating costs are applied to fewer products. If the production is higher, then overtime, or newer, less skilled labour is needed, again making the unit cost higher.

If labour is only required, the increase in output can be incremental, and much of shipyard production is like this. However, in many cases, the capacity can only be increased in steps, so if machines are required, then the capacity of an additional machine is fixed and may well be higher than is actually required. The step change in capacity will increase the unit cost.

A financial analysis of the proposed development is required. This will require estimates for the costs of development of the site. A survey of the proposed site will identify its suitability or otherwise, for all the facilities on the site. This will consider the ground conditions, potential need for piling large structures and load bearing

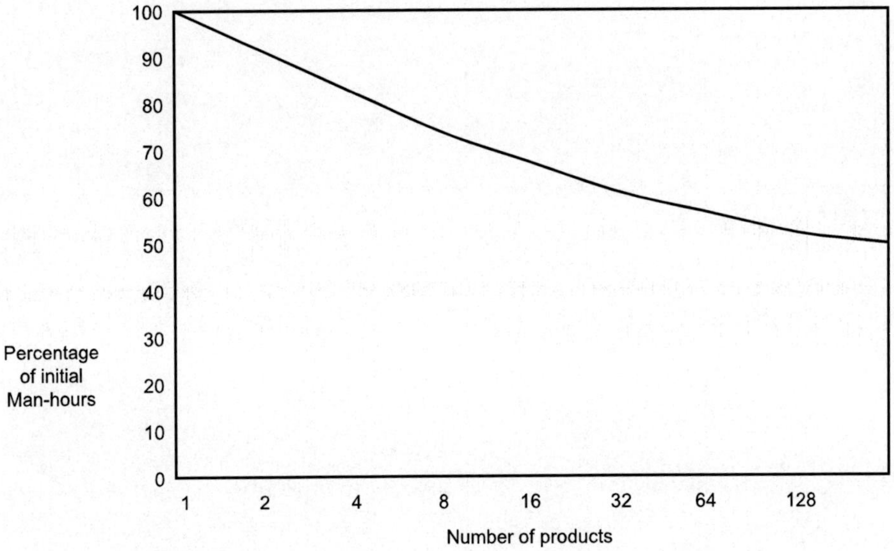

Fig. 5.3 Learning curve for ship production

areas and any contamination. There may be reclamation of land, restoration of land which has had previous industrial use, marine facilities such as a breakwater to avoid rough weather affecting the site.

Competitiveness has been defined as the ability to win and execute shipbuilding orders in open competition and stay in business. It is essential that a shipyard is developed so as to create the most cost-effective result. For each activity as it is designed, a trade off between expenditure on advanced equipment or on labour must be made.

Chapter 6
Ship Project Strategy

The best way to predict your future is to create it
—Peter Drucker

The strategy for the construction of a new ship requires solutions to a number of questions. For the initial strategy, the solutions are provided in outline, then they increase in detail once a contract is secured and the project develops. The first issue is to decide what is to be produced which is determined by the design activities. It is not the function of this book to discuss ship design in any detail, but a brief overview is offered here. More information on design for production is in Chap. 9.

The production of a ship depends on the generation of a large set of information. Historically, technical departments were concerned primarily with vehicle function, and the design work was limited to that. As a result most of the information needed for production was developed within the production departments, including the mould loft, pipe workshop and others.

Design is concerned with the definition of the ship and is carried out in several stages. Three are concerned with ship operation, described in this book as conceptual design to create an initial ship design to fulfil owner requirements, preliminary design to create a ship on which a contract can be based, and functional design, to ensure the ship receives owner and classification approval. These three are based primarily on systems, but in order to build the vehicle information is needed on interim products.

Transitional design is the translation of the design features from a system to interim product orientation. Although it is defined for convenience as a stage of design, in advanced organisations it starts at the earliest design stage. The later stages of detailed design are primarily to develop production information well in advance of the actual production operations as described in Chap. 9.

The second issue is to decide when the product is to be made, which includes the planning activities which determine the timetable for production. Planning includes design, information preparation and purchasing, as well as making of interim and final products. Planning has several stages, which parallel design. These are described in Chap. 10, and are strategic planning, which schedules the whole product, tactical

G. Bruce, *Shipbuilding Management*,
https://doi.org/10.1007/978-981-15-8975-1_6

planning, which schedules the planning units, then detailed planning and scheduling for work stations and individual interim products.

The third issue is to decide where the ship is to be made. This includes facilities engineering, described in the previous chapter. The level of definition varies considerably between different shipyards. This function may be left to production management or supervision to carry out at a late stage. However it is often a separate discipline in a more progressive company.

Where a company builds using relatively standard interim products, the work stations where they are to be produced will be engineered to give a close fit between design and production. This may lead to very specialised production areas such as panel lines. In some cases the where question is answered by choosing to sub contract some work, where it is highly specialised or there is insufficient capacity in the shipyard.

The fourth issue which is resource planning determines what is required for the product to be made. The resources which must be allocated include workers, equipment and portable tools. This activity in the longer-term may be carried out by specialist departments, especially in a larger company. For example facilities development as above, and personnel managed by a human resources department. The determination of resources in the short term and so for a single contract is an estimating and then a planning function.

What is required to construct a ship is also about materials procurement. The technical information includes a definition of all the bought in materials, equipment and services needed to produce the finished product. This information, along with a schedule defining when each item will be required in production, is the basis for procurement of all these items. The need for appropriate timing of procurement is set by the interim product approach to production. The procurement dates are based on a preliminary identification of the products as part of the initial ship project strategy.

Finally, the question of how the product will be made must be answered. This is for production engineering, and except in the case of new or substantial re-development of a shipyard this is often not a distinct function. It may be left to production management or supervisors, or be part of the planning function. The activity includes the breakdown of the final product into interim products, decisions on the processes to be used and decisions on the design detail to support production.

The key aspects of the pre-production activities in the project strategy can be regarded as modelling the production process in advance. The ship is broken down through an interim product hierarchy to individual piece parts. Production is then the process of reassembling these into the completed ship.

The objectives of a project strategy are to ensure that the risks inherent in any ship project are eliminated as far as possible. The strategy is initially the application of a shipyard overall manufacturing strategy to an individual contract. For a well-organised shipbuilder as discussed in Chap. 5 this will describe a product range which the shipyard wishes to build and a set of facilities designed to suit the chosen product mix. Within the facilities will be a set of production processes and methods also suitable for the products to be made. These will be supported by a technical function which responds to the needs of production and design. The operations of the shipyard will work to meet productivity and cost targets which will lead to success. Finally, administrative and other systems to support the strategy must also be in place.

The project strategy therefore starts with the existing manufacturing strategy, although if a new ship contract is for a different type then the strategy will also manage the variations which may be required. An important aspect is to ensure that all departments contribute to planning the project, so for example so the design function will take production requirements into account.

Apart from avoiding potential problems, the project strategy will also create opportunities to reduce the work content by the identification and use of suitable interim products. It will determine resource loadings and skills requirements, in the process identifying any shortfalls in capacity of facilities or labour, especially if there is a variation in the product mix. It will plan the design and purchasing of materials as well as production activities and in general provide a basis for all planning and project management.

Most importantly, the strategy will identify and resolve any problems before work begins, and communicate consistent information to all departments. So the product strategy is an agreed plan for the project, including design work, technical information, material management, production and testing. It is prepared before work starts, to identify and integrate all processes. It is a formal document, which gives statement of company intent, and which may not be amended without formal agreement. The strategy, without any confidential information, is often a requirement of the ship owner, as a document to be submitted with the shipyard bid, to give the owner confidence that if the contract is awarded the shipyard can fulfil it.

It has been argued that there are potential disadvantages of formal, written project strategies. Objections include the time and effort required to prepare the strategy. However the planning of the project is necessary if it is to be successful, so the effort will be spent in any case, to plan properly. Collating this into a formal strategy requires little extra work, and having the information in writing should ensure nothing is forgotten.

Another objection is that there must be provision to deal with unexpected delays or other problems. However the need to manage problems does not negate the need to plan properly. In fact the project strategy can anticipate, and often avoid the very problems which may cause delays through a careful risk analysis. And an objective of planning the project is to avoid changes to schedules which inevitably cause confusion and delay.

The strategy may be regarded as constraining flexibility and creativity. These are good during the project development phase, but once a production plan is committed, attempts to be creative will probably cause more problems than they solve. Good ideas are better left to the next project, with time available to adopt them.

The content of a project strategy will vary, and the precise contents will depend on the extent to which the new project differs from previous products. Typical contents can be summarised as first an outline specification of the ship, and especially any unusual technical features such as specialised cargo equipment. Then the contractual dates and constraints imposed by other on-going work in the company. Basic design and technical information for the ship is useful to inform all readers. There will also be information on major materials purchasing, especially for items with a long lead time which will need careful attention.

The work breakdown for the ship and the strategic planning are based on the initial design and are outlined in the strategy. Quality assurance requirements and a programme of tests and trials, again with key dates are part and finally human resources needs.

Any potential problems will have been discussed and solutions agreed as the strategy is developed. Almost by definition in a project as complex as a new ship, there will continue to be some uncertainty. This can be because elements of the project are uncertain, for example in the use of a novel material in a new application. Technical uncertainty, novel features and potential difficulties in production are common areas of risk for such projects. They typically affect the time and cost to complete the risky work. As a result, the time and effort needed to complete the project may also be uncertain.

Unless all the activities are entirely matters of routine, in which case we are no longer dealing with a project, the estimates which have been made will have a greater or lesser degree of uncertainty. An important element of the project planning process is the assessment of areas of uncertainty or risk in the project and as far as is possible taking steps to eliminate or at least reduce the risk.

The development of a project strategy is the mechanism used to manage the risk at the project definition stage. The uncertainty in duration of work tasks can be managed using network planning, where the timescale for a task can be assigned alternative times.

Where the ship to be constructed does have some novel features, a use of the strategy is to manage the process of change in production. There will be a need for change in the shipyard sooner or later, first because all products have life cycles. An extreme example is the traditional cargo liner, displaced rapidly by container ships in the 1970s. The container ships have continued to be developed and product changes are frequent, for example to open hatches and of course increases in size. A major change requires either some large adaptation of the shipyard or if that is not possible, a move to new products. The more specialised equipment there is in a shipyard, the more difficult it will usually be to change products.

Even for standard ships, over time the competition becomes more efficient. New competitors may emerge with advantages such as lower labour cost and as a result a shipyard must adapt to remain in the market. Again, if this is not possible a move to new products is required. Planned and controlled change is better than reaction to a crisis. If the shipyard management does not keep in close touch with product, technology and other changes, then it will eventually need to make large adaptations very quickly which is always difficult.

The project strategy provides a mechanism for managing change. The diverse elements of change are co-ordinated through the strategy including project planning, facilities development and usually most importantly productivity improvement. The commitment to a particular plan gives structure to the changes to be made by the management. Many shipyards will already have the information, but not in a consistent form. The strategy is a document which collates all the relevant information.

Although the shipyard strategy and project strategy are discussed here as separate aspects of the management, in practice they will be closely aligned and the pre-contract stage activities should merge into the actual project. Formal project management normally commences when a shipbuilding contract is likely to be agreed and ends when the ship has been delivered to the ship owner.

The project management procedure structure for a shipbuilding project is essentially set up for planning, budget, resources and technical specification. Although the focus is generally on technical issues, the strategy for a ship construction project will also be dependent on the contractual arrangements. General information on contracts and how they may be managed is discussed in Chap. 7. The contract must not contain any requirements placed on the shipyard that may be difficult to manage and achieve. Equally, the technical information for the ship and the project planning must be such that the contract can be fulfilled.

Alongside the shipyard project management, the shipowner will set up a parallel operation in order to verify that the new ship is designed and built in compliance with the shipbuilding contract and specification. To reflect this, the following section has been written as from the owner's viewpoint of the lifecycle of a contract. The owner or operator views the ship as part of a larger project which will provide a service to their customers requiring transport. It also helps when a management is able to understand the thinking of the other party to a contract. A shipyard should interpret the comments made from its own perspective. It is useful to bear in mind that the shipyard contractor may also be the 'owner' in respect of major contracts such as a sub-contracted hull, and is a customer for major engineering items.

Fewer shipowners employ a technical design staff than used to be the case, so the project may be managed by a design house or consultant. There is an inherent risk for the shipyard in this, as the third party manager may be less flexible than the owner.

An important part of the contract sets out the technical requirements for the ship which is to be constructed. These are a general specification, hull specification, machinery specification, electrical specification and a maker list. It will also specify which international maritime regulations and classification society rules will apply. There may also be questions of any charterer's requirements and owner's quality objectives.

The contract will intend to make sure that ships comply with approved plans and that the specified equipment and materials are used. All materials and components will need to be certified in accordance with the specifications so as to be reliable for the operation. It is important to beware vague phrases: such as "first class marine practice". These are open to interpretation and are potential aspects for disagreement. Specific requirements, ideally based on some internationally recognised standards, are preferred.

As soon as the project is expected to result in a contract, and before that is agreed and signed, some management activities are required. The customer's project manager will develop a contract management organisation parallel to that of the shipyard. This is to ensure that there is someone responsible who is available so that the shipyard can discuss issues with them. They will identify all contractually-anticipated communications between the client and contractor. They will review the ship specifications to remove any ambiguities and to assign responsibilities for managing problems.

As part of the management they will develop a specification for schedule reporting during the contract and review the specifications for tests and trials requirements. They will co-ordinate specifications with the contract plans and co-ordinate their agreement with specifications and plans. They will develop a spreadsheet or other register to manage the deliverables. They will also develop the terms and conditions which will apply to the contract.

In the early stage of a contract, after signature, there are further management activities for the customer or their manager. The first action is needed to check that the shipyard schedule of work includes all the expected items in the work scope.

The owner's manager will also need to arrange for the procurement of any owner-supplied equipment and materials. The parties to the contract should have a common understanding of the condition of any owner supplied items on their arrival at the shipyard. It is essential that the owner or their manager will apply the same diligence and commitment to this as the shipyard is expected to apply to the ship. Agreement over the date for delivery is essential so the owner must co-ordinate the equipment delivery schedule with the shipyard planning requirements. Delays or deficiencies in such equipment are also a potential source of problems. If equipment is late or incorrectly specified, the shipyard planning will have to be amended, but the owner may not want to accept the consequences as their fault.

It is also necessary to arrange and co-ordinate the owner's secondary contracts and support services. The arrangements for drawing reviews, including bills of materials are essential. Any owner comments must be supplied in a timely fashion, which should be specified in the contract. Failure to provide timely comments can be a source of friction, and potential delay.

The manager must ensure that the inspection staff, who may be contract only or third party employees, know the contract-defined standards to be applied. These must be agreed in the contract and preferably be recognised industry standards.

Once the contract is in progress there are some on-going management activities required for the owner's manager, of which the most important are mentioned here. The manager must keep oversight of communications between the classification society and the shipyard. A commonly preferred situation is to have the owner's manager copied into all classification communications sent or received by the shipyard within one or two days of transmission or receipt. The same arrangement should apply for regulatory communications, but getting timely response from the regulatory bodies is often more difficult.

Updates of the critical path network schedule need to be reviewed where there are any new or revised activities agreed between the owner and the shipyard. Progress has to be monitored against the shipyard schedule at joint progress meetings. Any commitments made at progress meetings should be based on adequate research as to the probability of achievement. The contract documents should be examined first to check if they provide any answers to any questions raised.

Progress payments will have to be made. These should be based on measurable progress in design, procurement, physical construction and also completion of non-hardware deliverables such as reports, operating manuals and as-built drawings.

Carrying out inspections and witnessing testing of components and systems is an on-going requirement. The purpose of owner inspections must not be confused with inspections by classification or regulatory bodies or the shipyard's own quality assurance function. Each has its own purpose, for the shipyard to check their interim products and that their suppliers are meeting contract terms and that the supply will operate properly. For the classification society and other regulators, the purpose is to check compliance with standards. For the owner, the purpose is to ensure progress.

There are also management activities which are required from time to time, as and when they arise. The first is to review the technical and cost aspects of proposed changes. The most severe problems arise when the owner requests a change without a sufficiently clear definition of the change, giving rise to a potential dispute when the ambiguities in the original request are interpreted differently from the owner's intention. Another potential source of dispute is any disparity between the owner's estimate of the cost of a change and that of the shipyard.

It is necessary to analyse the impact on the schedule and extensions of any proposed changes. The justification for an extension may be linked to the scale of any change or the total number of changes introduced. Even a contract clause which specifies there should be no schedule impact may have to be relaxed in later negotiations if the total number of changes introduced by the owner cannot be handled without some impact on the delivery date.

It may be required to negotiate contract changes, but this must only be done if it is unavoidable. The key to managing contract changes is rigorous maintenance of up-to-date contract documentation (covering the price, technical specifications, drawings and delivery date) so that the cumulative effect of changes in price, weight, stability, performance and any other aspect of the ship gives no unpleasant surprises. The owner should enter into negotiations with a combination of targets, with a maximum acceptable cost and effect on the schedule in mind before starting. Directing a shipyard to perform a change before the change has been agreed as to the cost and schedule impact should be avoided. The contract must then be updated to reflect changes.

The manager has to monitor any potential delays asked for by the shipyard, and force majeure delays. The wording of force majeure claims should be checked when a request for delay is received. For example, specifying delays due to "unusually severe weather" as acceptable implies that delays due to "usual severe weather" are not. Supplier or subcontractor delays may not be allowable by the contract and even if they are allowable, the shipyard must demonstrate that it took timely action to mitigate the consequences of such delay. Delays due to late supply of owner furnished equipment or information should be tested for cost and or schedule impact. The shipyard may need to carry out some rework and this is likely to have an effect on cost and schedule.

Towards the end of the contract, and in the post delivery period, further activities are needed. The owner's manager will develop inspection delivery reports for nearly-complete work to ensure no impact will occur on the construction schedule or the eventual performance of the ship. The manager will also approve the agendas for and then carry out oversight of all tests and trials. These are limited to those anticipated in the contract or by documented industry standards. A greater scope of tests and trials will require a mutually-agreed contract change.

The manager will prepare acceptance arrangements for inspections, tests, trials and reports. Acceptances provide a potential source of friction between owner and shipyard due to the approaching end of the contract and the shipyard's desire to meet progress payment points. Tests and trials often require a surprising amount of time and effort to prepare for and carry out. Correction of deficiencies that prevent tests and trials from proceeding may be implemented as short-term, expedient quick-fixes. However the long-term correction may never be implemented unless care is taken.

Compartments are closed out after completion of all work. The owner and shipyard should negotiate a reasonable programme of presenting compartments for close-out inspection to avoid owner's staff overload. There will also be a review of non-hardware deliverables prepared by the shipyard.

The manager will have to approve spare parts lists and arrange for their timely arrival. The manager will also review 'as-built' drawings for conformity to changes. Another need is to develop documentation for delivery, including three lists. These are outstanding inspection reports agreed not to be closed out, but for which a price adjustment will be made. Also inspection reports that the shipyard agrees to correct within the first half of the warranty period, and against which the owner has withheld some payment. Inspection reports in dispute, where some payment may be withheld against these items subject to post-delivery negotiation or arbitration or litigation are listed as well.

Management will identify warranty items according to contract procedures. Warranty items should not be confused with outstanding deficiencies. The contract should set out the procedures, rights and responsibilities of each party in respect of the timing, notifications, correction of warranty defects, costs and withheld sums. The burden is on the owner to prove that a claim is a valid warranty matter and is handled in a timely basis.

As a final comment, both parties to a contract must be active participants during performance of the project because being passive is likely to cost money, whereas being active can save both costs.

The owner's project organisation will to some extent mirror the shipyard equivalent. There will be a team who have clearly defined responsibilities, tasks and reporting requirements to perform the work. As has been mentioned these personnel may well be from third party organisations, or may be recruited specifically for the ship contract in hand. The most important is a project manager who is the counterpart of the shipyard project manager.

He should have authority to act on behalf of the ship-owner and overall responsibility for ensuring that the work is carried out by qualified and experienced personnel in a cost-effective way. He also monitors and ensures that the work is verified, documented and carried out at the agreed time and price. He may have other duties for the owner, for example with other shipyards building ships or part of fleet management. These should not be allowed to interfere with the smooth running of the project.

A project supervisor will operate permanently in the shipyard to lead the owner's project team, to ensure that the specified quality and optimum use of resources is obtained. One of these will be a project co-ordinator who will be responsible for the arrangement of approval work and regular progress reporting to the ship-owner. There will also be a team of inspectors, one of whom will be a site chief inspector. The chief inspector will usually be appointed prior to commencement of construction, bearing in mind again they may not be permanent employees of the owner. The chief inspector will act as the ship-owner's representative on-site and official contact between the shipbuilder, project organisation and the ship-owner to ensure satisfactory progress.

The size of the project team will depend on several factors, including the size and complexity of the ship being built. The location of the shipyard, the level of experience they have of the ship type and often the previous relationship between owner and shipyard will also be considerations.

The shipowner, through the representatives, will follow the project through a number of stages which mirror the stages described in the shipbuilding activity map. At each stage the owner's representative and the shipyard will need to liaise closely so that the information on progress of the contract is clear and agreed between the parties.

First is the set up of the project, starting prior to contract agreement and focussed on the project planning to identify the activities and resources required to obtain the desired quality. This will include the procedures which will be used to control and execute the work. Further, there is a need to have factual information about the project to ensure that the owner's project organization is properly informed about all the major aspects of the work to be done.

This will take much of the information from the shipyard project build strategy and the associated preliminary design work. A project procedure manual is usually prepared at this stage to set out what activities will need to be carried out at each stage of the work. This will inform the owner project team as it is assembled, considering some members may not be appointed until work has commenced and there is a specific need for their participation.

The owner's project manager will create a project plan which defines the division of work, individual responsibilities and budgets. This plan will include budgeting, the future project meetings between the shipyard and owner, management of plan approval, management of site inspections.

The owner's team will receive written work assignment and information from the plan to inform them of the scope of work, starting, termination and reporting procedures. The project supervisor will need to discuss and clarify each main step and ensure common understanding of the task to be performed. This will be required for the owner's team and the shipyard project manager.

Plan approval includes all the drawings and documents used in the design and construction of the ship. The purpose of plan approval is to evaluate the plans to ensure they comply with the shipbuilding contract and ship specifications, the relevant maritime authorities' requirements, the classification society regulations, any charterer's requirements and the ship-owner's intended operational patterns.

Evaluation should consider whether proper and cost-effective maintenance and repair work can be made throughout the ship's life time. Procedures include noting that conditions and limitations of the approval may be presented on the plan and in the accompanying letter in the form of comments. All comments must be clearly shown and generally signed and date stamped.

Where comments have been made, the shipbuilder will need to submit revised plans to the owner's plan approval organisation and to the shipyard site office for verification that comments have been implemented on the final plans. If there are many comments or modification, resubmission of the subject plan may be requested.

Design changes may be requested by the ship-owner during the plan approval process which can affect the contract price and the delivery time. Ideally, the changes will be submitted before any work commitment and there is often a cut-off date after which any design changes can be refused by the shipyard because of their effect on the schedule. This may be not only on the current contract but also on other concurrent work in the shipyard, of which the owner will not be fully aware. Any changes must be approved by the owner's project manager and the shipbuilder and must be formally recorded.

The selection of makers for ship's equipment is important. The aim for the owner is to select equipment, machinery, materials and components that have been evaluated and found to be of high quality with recorded good reliability, designed for cost-effective operation and future planned maintenance routines. The proposed items and materials should be verified for their functionality, reliability and maintenance characteristics.

Reliability is a requirement of classification rules and is arranged in the way that a single failure in a sub-component will not cause loss of main function for a long period of time, defined by the time required to restore the function. Critical components require a redundant component to maintain service, main components need their function to be re-established within a specified number of minutes and sub component within a specified number of hours.

Operational experiences with components and materials from other ships are used as a basis for the availability calculations and makers evaluation. Owner's usually avoid novel and non-conventional components and materials unless an in depth evaluation of the reliability and availability can be performed.

The components should be able to be maintained or repaired in a cost effective way. An evaluation on the availability of service engineers and spare parts for all main components should be prepared by the shipbuilder or maker covering the availability of service engineers at main trading ports planned for the ship, the cost of service engineers and availability of spare parts.

The instruction and maintenance manuals provided by the makers should be evaluated based on the content of operational instructions, the coverage of maintenance and repair instructions, recommended spare parts, including a reference system for purchase and planned maintenance system information.

The site inspection period will generally be from the first steel cutting until the delivery of the ship. The inspection scope of work, depending on the type of ship, covers hull construction, surface preparation, application of anti-corrosive coating, machinery installation, electrical and automation installation, cargo system installation, the outfitting and superstructure.

All inspections have to be properly planned, documented, recorded and signed by the site chief inspector, or attending inspectors and also the shipbuilder's representative. The site chief inspector must plan and coordinate the site inspection tasks in accordance with shipyard production schedule and agreed quality plan. Normally inspection includes scheduled inspections, unscheduled inspections, inspections of incoming materials and components, inspections at subcontractor premises and on site inspection.

On completion of an inspection or test, the relevant inspector must inform the shipbuilder of any observed non-compliances, defects, deficiencies and any conditions which need improvement. Comments will normally fall into one of the following three categories: Those that must be corrected before work in the area inspected is allowed to proceed, comments that need to be corrected before final acceptance of the task is given and any instances of non-compliance with requirements.

The shipbuilder will implement procedures for the identification and handling of comments. For control and follow up of comments, the following procedure is normally adopted. First, comments are always given in writing with reference to work failure, standards, specification or rules and regulations. The ship builder generally corrects or responds to the comments within seven days. Where the shipbuilder does not correct or respond to the comments, the site chief inspector will issue a reminder to the shipbuilder.

A non compliance request requires corrective action to be taken which must be approved by the owner's project manager who is authorized to permit deviations from building specification in certain circumstances. The shipbuilder's corrective action must analyse the cause of the defect or other problem, plan for corrective actions, with deadlines for implementation, then actual implementation and follow up inspection.

The scope of tests and trials generally includes function tests of machinery, equipment and systems, tank structural and tightness tests, pipe structural and tightness tests. These are carried out as early as possible as work proceeds, which benefits both owner and shipyard. Then, docking tests and measurements, an inclining experiment and finally a sea trial are carried out.

The procedure for approval of any test describes the test procedure to be carried out, a reference to appropriate standards and acceptance criteria and the necessary performance data as stated in the ship specification.

The owner supervisory team must ensure that shipbuilder has implemented the systems required for quality control and necessary approvals from external organisations. They also ensure that there are satisfactory, procedures for calibration of any measuring equipment and standard procedures for all tests which must be carried out.

The supervision reporting system should be organised in a systematic manner so as to document all the work carried out, in particular the compliance of ship assembly, construction and installation with the approved plans. The site inspection supervision reporting system must enable the identification of all inspections and tests carried out, the correct performance of inspection work, all recorded comments, non-compliances and all corrective actions that have been implemented.

The reporting system can be structured in three levels, an inspection or checklist report for each task, inspection progress reports periodically and an inspection summary report to the ship-owner at specified intervals.

The inspection or checklist report is primarily a tool for the owner's inspectors to specify the main steps, areas of importance and the parameters which are to be recorded from inspection, test and trials. The inspection progress report records the inspection areas and items, enabling evaluation and preparation of actual inspection work progress. Therefore the inspection progress reports normally include a list of items inspected, the inspection date and the status of inspected items and a summary of comments.

At the end of each specified period, often a month, the project supervisor evaluates the progress and prepares a report to the ship-owner covering the status of plan approval, maker selection, the status and progress of inspections on site and any observations and corrective actions. Important for the owner are any matters affecting progress and the ship delivery schedule, any milestones achieved during reporting period which may trigger a progress payment, and matters which can affect the contract price and any matters of dispute between shipyard and owner's team.

Considering the large quantities of plans, documents and correspondence which will have accumulated during a shipbuilding project, a robust computer system should be implemented for recording and control. This will keep track of drawings, comments given on each plan and comments implemented, giving an accurate real time status of plan approval. This is vital for ensuring that all comments raised during the plan approval stage are implemented within the construction drawings and actually built into the vessel.

By implementing a project procedure manual, the management of shipbuilding projects can be executed with stringent quality control procedures. This combined with the application of a regular audit programme to assess all aspects of management work, enables a standard system to be established. The benefits to the ship-owner include cost effectiveness and properly documented supervision carried out in line with a well established industry standard. Such management of shipbuilding projects not only ensures that ships are built to highest level of safety, quality and reliability in accordance with the specification but assists the shipbuilder in delivering ships on time and within budget by identifying and proposing solutions to problems before they affect cost and delivery schedule.

A close cooperation between the shipyard and the owner's representative will usually result in a successful contract which follows the strategy prepared at the outset.

Chapter 7
Commercial Activities

All courses of action are risky, so prudence is not avoiding danger (you can't) but calculating risk and acting decisively
—Niccolo Machiavelli

While most of the activities in shipbuilding are technical, including the shipyard arrangement, production technology and planning, it has to be remembered that the ultimate objective of the shipyard is to make profits and so stay in business. So the first requirement for a shipyard management is to obtain profitable contracts. As a result, the commercial activities are fundamental. They also have to be closely aligned with the technical activities so the two are mutually supportive.

The commercial function is generally responsible for dealing with the customer and for financial elements of a contract. So managing customer relations is a good point to start to consider the commercial activities. Game theory provides a useful way of looking at the potential situations which arise between customer and contractor, or between contractor and supplier. Game theory offers a useful means of considering situations in business relations. Various "games" can be used and two are outlined here, zero sum games and win–win games. The ideas are useful in considering "transactions" in projects and in judging how a shipyard can manage relations with both customers and suppliers.

In a zero sum game, the outcome of a transaction is such that there is a fixed value, so in a contract if one party gains then the other loses. This is adversarial, and sadly is often found in shipbuilding contracts. The likely outcome is a lack of co-operation and potential conflict between the parties. A possible extreme outcome is legal action, in which case the fixed benefit is reduced and so both parties may lose because litigation is very expensive and the legal system is itself often adversarial. Zero-sum is something to be avoided if possible in most cases. Although it is often assumed that the costs involved in a ship project are fixed, often reductions can be made so the value of the transaction increases.

A win–win game is another possibility which uses the idea that outcome of an activity allows for increased benefits. Again an example is a contract between two parties, where with co-operation there can be cost savings or increased benefits. If the parties can agree how the savings or increased benefits may be shared then there

G. Bruce, *Shipbuilding Management*,
https://doi.org/10.1007/978-981-15-8975-1_7

is can be mutual benefit. The win–win game is the basis of many contract forms, for example alliances and cost-sharing arrangements. Both can lead to a creative search for opportunities to save costs.

Examples of both games can be found in the industry, from the symbiotic relationships between large passenger shipbuilders and their main interior suppliers to entirely adversarial attitudes of shipyards to small suppliers. While a cooperative approach to a contract is ideal, it is also dependent on the power relationships between parties to a contract. Their relative status in a project is important. If one is dominant, then a zero sum is the likely result.

In summary there is a lot more to game theory than presented here. However what is identified here is the need to consider the relationship between parties to a project. Ideally a cooperative project between equal partners is the best way for a ship contract to be progressed. Sadly in the marine world this is not always possible.

The relations with customers start with the shipbuilding market. Market research is fundamental to successful shipbuilding so close contact with the shipowners, operators and often shippers is crucial. This provides input to preliminary ship design activities, and then to production strategy.

Gaining publicity for the shipyard and maintaining a high market profile is needed alongside the direct marketing to potential clients. Trade exhibitions offer an outlet for this. They can be expensive, in terms of the cost of booking space, travel, accommodation and equipment needed. There is also the cost and the time of commercial staff and some technical staff to attend. On the other hand, some of the exhibitions provide an excellent opportunity to meet useful people in a good atmosphere. It is also useful to meet other exhibitors, shipyards and suppliers, and gain further market intelligence informally.

Trade publications provide a useful outlet for publicity material about the shipyard capabilities and contracts. Provided the material is not too heavily based on advertising, publications are usually very much in need of new articles. Often they are even happier to provide a platform if the shipyard also buys advertising space. Research is needed to be sure the publication is the best for a shipyard's target market.

A more long term and more difficult route is to seek to publish in the journals and transactions of marine engineering and other professional organisations. Some papers for transactions may need to be peer reviewed, which can be a slow process, although it also lends a lot of credibility on publication of the work. Journal articles are easier, and faster to get into print or make available on line.

Trade associations are usually national organisations which represent an industry to government and the market. They can offer representation to the industry as a whole, and offer practical assistance, for example by organising a joint industry presence at exhibitions and other maritime events. Even if the industry is a small part of a national economy, the trade association will give it more visibility and better access to, for example, government.

A trade association may improve the opportunities for grants and other government support to attend trade events, such as exhibitions and may even be able to obtain sponsorship for focussed trade missions. Government forums may exist such as joint industry-government committees where the interests of the industry can be represented and promoted.

Once a potential contract to build a ship is found, the process of contract management becomes the main focus of the commercial activity. Some outline principles are offered here to provide a basis for managing contracts on major projects of which ship construction is a good example. They must be seen as support for the application of project management best practice in order to maximise the probability of completing a project successfully. Completion on time, on budget and in conformance with the specification is a good definition of success for a contract. These principles are presented in two parts, first general principles affecting all contracts and second specific issues affecting the shipbuilding industry.

It is worth remembering that the shipbuilder can be involved in contracts as both a customer and a supplier simultaneously for a single vessel, as supplier to the end user or owner and as customer of the company's subcontractors.

As a general statement, a good contract can contribute to a successful project. A good contract will motivate all parties to achieve the project objectives. It will govern how the members of the supplier organisation will behave with the customer and other project team members who can include their own suppliers and subcontractors.

If a project is fully defined and all risks have been eliminated, then both customer and supplier will know exactly what is to be delivered and that there is certainty this will be achieved. In that case a simple purchase order is all that is required. However that is unrealistic in a shipbuilding project, and the project would not require much in the way of project management. So the contract is largely needed to record the obligations of both parties, depending on the uncertainties and scale of the project. It allocates residual risks in the most appropriate way and states the policy and principles which will govern the project. Contract terms will usually reflect the past experiences of customer and builder, and so their initial approaches may be very different. Negotiations are required to move to a mutually acceptable position.

Planning a contract is a project in itself, as there are all the characteristics of a project. There are objectives, stakeholders perhaps with varying requirements, a work breakdown structure, risks and a programme to develop the contract. Its organisation and the allocation of responsibilities must be decided.

In pre-contract discussions, the available types of contract can be considered, with the risks entailed in a contract and the terms of payment. What contract documents are required, how the contract will be administered and how the contract closure will be managed must be agreed. For a large contract, the customer and supplier should agree some common objectives. The responsibilities of customer and supplier will be set out, and key programme dates agreed. These will initially be the completion of negotiations, then completion of the work and sections of the work, with the stages of completion usually leading to stage payments.

The contract exists because some risk remains, so decisions are needed on who will manage each risk, how variations in scope will be controlled and how disputes will be avoided. In short this is to ensure that any problems of contract management will be anticipated.

The contract structure is important. It will set out the programme with approximate timescales to enable a contract plan to be produced. This must be linked to key project plan dates. It is essential to allow time for checking that the supplier's offer is consistent with the customer's requirements both technically and commercially. If differences exist the contract should include agreement on how they will be resolved.

Some organisational considerations are relevant. In general the customer organisation should be small and only consider technical compliance and payment of invoices in accordance with the contract. The customer's contract manager should manage changes to scope of work and then consider cost and time with the contractor. It is important to remember that it is the contractor's job to execute the work and the customer's role is primarily to manage the contract which is a legal and commercial understanding. The customer must avoid becoming the contractor's project manager or technical expert, so the responsibilities should be clearly defined for all contributors to a project.

Pre-contract discussions are valuable for understanding both customer and contractor requirements and exchanging options not previously considered. They offer an opportunity to identify what information is available and the contractor capabilities. In a competitive tender, all contractors should be given the same information. The customer project manager and ideally also the potential contractor project manager should be present at all pre-contract negotiations to ensure the contract records the true intentions of the parties. They keep a day-by-day record of all discussions and decisions and the reasons for them. The contract must state whether documents exchanged at this time are part of the contract. This provides continuity for a project which may last for years.

At the end of these discussions the customer must define what is wanted in terms of cost, quality, delivery, safety and any other needs. The customer can then choose the contractor best capable and motivated to deliver their needs. The choice of contract type depends on the responsibilities and risks associated the project. It also depends on the capacity, capability, quality and motivation of suppliers, the relative importance of quality, safety, time and cost and the ability of the customer to manage a comprehensive contract. It is also very dependent on the remaining level of uncertainty about the project scope at the outset.

Good contract management starts at the beginning with good planning and understanding of the risks for a project. Managing the things that could go wrong is just as important as progressing what is going well. Allocation and management of risk in accordance with basic principles will reduce frequency of costly and time consuming disputes.

Risk management is a very important part of the project, and affects the contract. Having established that a risk exists based on the project manager's experience and a formal register, an appropriate strategy is needed to avoid, or transfer the risk. At worst to mitigate and then insure against the risk occurring. Whether insurance is available depends on whether is it quantifiable in monetary terms. It must relate to "pure risks" where the only outcome is harmful and a sufficiently large number of independent risks must exist in a market so the insurer is able to spread his risk. The loss if it occurs must be unintentional.

The customer can mitigate risk in a review and revision of initial documents, assessment of potential liabilities and a check on realistic contract performance times. A reasonable budget contingency should be made to recognise indeterminate elements in the project. During negotiations ideally both parties should seek to define the "best deal", and make joint efforts to obtain it. There should be planned communication meetings and informative correspondence with decision-making delegated to appropriate levels. The contractor's project manager must be carefully selected and empowered to act directly in most circumstances.

Pre-negotiation risks may include credit worthiness, prior dealings and references. In the contract negotiation phase the terms of payment should be chosen to motivate the supplier to achieve the project's objectives. Fixed price contracts are often popular, but have risks, such as dealing with technology new to the supplier or managing different suppliers. Alternative terms of payment to fixed prices are available.

In cost-reimbursement the customer pays for authorised work, the supplier can be chosen before the work scope is well-defined and risks are only paid for if they affect the work. The amount and speed of work can be varied greatly, the supplier is not responsible for the economic use of resources and there is greater flexibility for risk allocation. However there is a serious danger to the customer of cost increases. Competitive bids can be invited by asking for rates per hour or day, or for categories of staff, to be specified, more like a shiprepair project.

Customer and supplier can agree a target cost for a risky project and agree that both will share savings of final cost compared with target. Then both parties have an incentive to limit cost and such contracts are often popular for military ships.

Most engineering, construction, fabrication or larger procurement contracts include a term for retention money. In this a fixed percentage of the total payment will be retained by the customer. There can be terms for reduced payments to the supplier if work is completed late, or if equipment achieves less than specified performance. These liquidated damages are intended to motivate the supplier not to fail. Some experienced customers include payment of a bonus for completion on time and to specification, including safety in installation and construction work.

Contract documents should define clearly the scope and performance requirements. Constraints should be specified in terms of cost, time and risk allocation. Some specific management issues are usually included in a contract. Projects are executed with three elements to be considered, these are technical correctness, progress and commercial issues.

Technical issues include an audit trail of the materials and goods which are used, performance measures, periodic tests for work completed, operational requirements, asset maintenance, a method of costing technical changes accurately and establishing intellectual property rights.

Progress issues include the project execution plan, monitoring the plan, planning for and analysing the possible risks, establishing a reporting structure, a process for reviewing monitoring and updating the plan and monitoring the risk.

Commercial issues are that payment should be in line with progress, linked to stage performance tests and documentation requirements. It is important to establish joint sharing of any improvements. All the contract management team should be fully conversant with all terms and conditions. A well developed payment plan schedule should allow for at least 15 or 20% retention to cover the contract closure and delivery and then the warranty period.

In practice, market conditions, new technology, design constraints or other unexpected events may force a variation to the original project scope and price. Procedures should be in place for managing variations, with instigation, evaluation and payment. Variation control by both customer and supplier should operate procedures to control significant variations. The customer should be kept fully aware of current and anticipated final effect of variations in budget and programme. Procedures should allow proposed variations to be evaluated for time and cost and to be agreed preferably before a variation is authorised.

Disruption caused by disputes can be avoided by selecting contract procedures that reduce the risk of those disputes occurring and by using efficient and equitable methods for resolving disputes that do occur. Alternative dispute resolution, of which an example is conciliation, is better and far less expensive than legal action. The principles for reducing the incidence of disputes are to use equitable risk sharing through choices of appropriate contract strategy. Tendering procedures should avoid or at least minimise the supplier relying on claims for additional costs to achieve a profitable project. Contract terms should be able to obtain agreement quickly on any additional time or cost when a variation occurs. Using contract conditions which are clear and precise in terms of how risks are allocated, and procedures to be followed if they occur are important to avoid disputes. The contract conditions should also motivate the parties to identify problems early and collaborate in overcoming them.

In closing the contract, the technical requirements are to have certified completion of scope of work, including variations and performance analysis after a period of operation during the warranty period. All auditable documents must be agreed and signed, technical information complete and passed over, including operating instructions, as-built drawings and test records. What triggers the warranty period, and when it starts are agreed, residual materials are transferred to stock or disposed of and customer-owned property is accounted for and properly disposed of.

Commercial questions to answer are whether all bills have been paid, have all outstanding disputes been handled satisfactorily, are the terms and conditions of the warranty understood, is the close out report agreed and are residual bonds and credits cancelled? Any claims and disputes must be resolved or recorded, potential claims and disputes eliminated, fees or royalties due under licence agreements paid, internal order numbers closed so that no further costs are allocated to the project and obligations with respect to offset arrangements fulfilled.

There is usually a warranty period. Technical and operational requirements include after sales service and mechanisms for triggering a warranty claim. Commercial issues are terms and conditions of after sales service issues and reviewing warranty periods.

At the early stages of a ship project, a letter of intent may be prepared, which outlines the buyer's wish for a ship. While not binding, as there will be conditions, the letter is intended to give both parties to the potential contract some confidence before they both commit to pre-contract costs. These can be substantial and in some cases an initial design contract might be awarded to a builder to cover the costs. Such a contract will be included in the main contract if the project proceeds.

The shipbuilding contract is often based on a standard form. These have been produced by various organisations, for example the Shipbuilders' Association of Japan (SAJ) and the Association of Western European Shipbuilders, later the Confederation of European Shipyard Associations (CESA). The general contents are always similar, but the details can vary considerably according to the specific contract circumstances.

The most widely used is the SAJ form, not surprisingly because Japan was the leading shipbuilder and the form was adopted by Korea and China as they developed their shipbuilding industries. These three nations dominate shipbuilding now. The form was drafted by shipbuilders so is often thought to favour the shipyard in a contract. Many of the amendments to the form have been made by shipowners to make the contract more equitable. The Baltic and International Maritime Council (BIMCO) produced an alternative form of shipbuilding contract, which is seen by the shipbuilders as more favourable to the owners.

The contract starts by stating the parties to the contract, with details of the customer and contractor, then the date and place of the contract. There may be several parties to the contract who will all be noted. The customer may be a part of a larger organisation, as may be the contractor, in which case the contract may include which one of perhaps several shipyards will construct the ship. All points of contact should be listed. There may be further shipyard information, as possibly the specific shipyard will be determined later for a large shipyard group.

Next there is a description of the ship to be built. This is basically a summary of the main characteristics, and will be supported later by a detailed specification, associated drawings and other technical information. The contract will note as a minimum the ship type, its main dimensions and cargo capacity, the maximum and service speeds for the ship. The main engine, usually with the maker, power rating for maximum and continuous operation and the fuel consumption under specific conditions will be specified. Any other specific requirements for the ship type, for example cargo loading speed will be listed. Further details will include the shipyard building number or hull number.

The contract price and the appropriate currency will be specified, with the terms of payment, generally requiring progress payments according to work completed, and the dates for those payments. Bank details are also required. Importantly the delivery date and intended payment dates will be specified.

The flag state under which the ship will operate, and therefore the national regulations which will apply during construction, also the classification society for construction, which may change after delivery.

There is always a possibility of the ship not being able to perform as specified, in which case there will usually be penalties. Also the builder may not complete the ship according to the contract schedule. Further details must therefore include the actions to be taken if there are failures. These can be failure to achieve contract speed, with incremental payment reductions up to a maximum figure. Another possibility is a failure to achieve cargo capacity, again with incremental payment reductions. The ship may have excessive fuel consumption compared to the specification.

If the ship is late, the failure to deliver on time will attract a daily payment reduction. Any other specified ship requirements which if not met will reduce the operating capability of the ship may also be in the contract.

Guarantees will be required, from the shipbuilder that the ship will perform, other than the failures above, as required. The buyer will guarantee to make payments as in the contract schedule, within specified time periods. A guarantee period after delivery, usually twelve months, during which time the builder will rectify specified defects is usual. There may be additional guarantees for some ships, and a guarantee engineer may be supplied to ensure performance.

The contract should be defined to avoid disputes, but it will be necessary to specify which law will apply if there are disputes and what alternative procedures for resolution will be available to both parties, short of going to court.

In some cases the owner may be interested to purchase further, similar ships and so there may be options for such purchases. The contract will identify the proposed prices and delivery dates, and also the dates by which options must be confirmed. The shipyard cannot have options undecided for too long if alternative contracts are being negotiated.

Both builder and owner will need to provide guarantors. This may be a parent company or a bank for example.

There is scope for further contract items to be agreed according to the needs of a specific ship.

The contract will then have a large number of clauses, which define fully the information outlined above. These must be very clear and unambiguous, so that any possibility of later dispute is avoided. The starting clause may specify that the builder is obliged to deliver the ship exactly as specified and on the correct date. It will also mention quality and it is essential that there is reference to specific shipbuilding standards. Any phrases such as building "to good international shipbuilding practice" should be avoided because without a detailed and specific reference there is immediate scope for later argument.

Cost estimating is of importance for the commercial activities of a shipyard. The cost of a ship or other large, made-to-order product must be estimated, simply to determine whether the construction of a ship is a viable project. This is necessary both for the customer, the owner or manager for the ship, and the potential builder. An initial estimate of ship cost is needed when the ship is no more than a potential product to be part of the shipyard strategy. The estimate determines the cost of building a ship, in a specific shipyard and at a specific time. It is an important element of the commercial activity, with the initial estimate of cost a key filter in selecting which of all the enquiries that the shipyard receives are worth pursuing.

The price for the ship is largely set by the shipbuilding market which is affected by supply and demand in the shipping market. This then determines the demand for ships, also taking into account the supply of shipbuilding capacity. The capacity is influenced by possible incentives from governments, since a lot of shipbuilding capacity world-wide exists because of government intervention, both to create a capacity and to maintain it. Incentives can be offered to shipbuilders and to owners, and may not be transparent because subsidies are against trade rules in many cases.

There are also exchange rate fluctuations over time so the value of an international project is likely to vary. Allowances may be needed for potential variations in the exchange rates, for general cost inflation and movements in material prices. Occasionally the costs of financing may also need to be taken into consideration.

There is a degree of risk in any estimate because of the above. Importantly, the actual cost of constructing a ship may vary considerably from the estimate and is likely to rise considerably for a prototype project. In fact the final cost of a ship is only known after it is complete, and even then will depend on the accuracy of accounting assumptions and data recording. As a result there are often commercial issues to be taken into account, which may involve accepting a risk that the estimate available when needed for customer negotiations will not be reliable.

There are some influences on managing the cost and price of a ship. The shipyard's experience is one because if there is limited or even no experience of a ship type there is very likely to be an underestimate of the cost of construction. A number of small shipyards have taken on contracts to build specialised ships, such as those for scientific research. These have difficult specifications which require very low noise and vibration characteristics, often developed by consultants with limited knowledge of ships and the marine operating environment. The result has been losses and in a few cases the shipyard going out of business as a result. So a balance must be found between the commercial imperative to secure a contract and the ability of a shipyard to fulfil all the details of it.

Prior to the contract, there is often very limited design information on which an estimate can be based. The development of at least an order of magnitude cost is required. There is a need to produce a credible price for potential customers so that the shipyard commercial function can decide whether the ship is viable as a future product. There may be discussions with potential customers and some price indication will be required before these can proceed. In some cases there is a close relationship between owner and shipyard, for example for large passenger ships, for military ships and for many classes of specialised, non-cargo ships.

At this point where the ship is just an idea, the order of magnitude cost will be used to review concepts, from which the most promising can be selected for further development. The accuracy at this stage is likely to be plus or minus 20% from the initial estimate. This is significant, but will be adequate for initial product selection and will be refined if the product is selected for further study. This estimate will also be used by the shipyard if there may be investment in the facilities.

An approximate early indication of cost can be developed from main characteristics of the ship. For example, it can be based on the cargo capacity of a ship. Given records of past contracts a formula can be developed which relates the cost of a ship to the capacity. If the proposed new product is similar to previous ships, perhaps with just an increase in size, or with some novel equipment, then this is a useful starting point for the cost.

If an initial design is to be developed, because it seems to be of interest to potential customers then a shipyard design study will be carried out. This is where there is a potential project and the design can be taken to a stage where a contract may be signed. How far the design is to be developed is variable. Typically at this stage of the design development, the estimated costs will be plus or minus 10–15% of a likely outcome. A large degree of commercial judgement may be required in deciding to accept a contract.

In some cases the design will be prepared by a consultant to the shipowner, if there are very specific requirements. Such a design may be more complete and will be sent to prospective builders for a quotation. In other cases, there may be a joint development between a shipyard and regular customer, so the design development may continue after agreement to proceed.

Overall the commercial function has to manage the financial risks in a contract, always in close collaboration with the technical and other functions to achieve the best result for the shipyard.

Chapter 8
Materials Management

Don't let the buyer beware, let the seller be honest
—Isaac Asimov

This chapter describes a number of aspects of materials management. The first aspect is the importance of purchasing to the ship construction and how this function is linked to the other activities. It then considers the need to manage the relationships with suppliers, ideally through co-operation with them. There is a need to expedite deliveries, then once materials are in the shipyard, there is a need to minimise inventory and to package materials to suit production requirements. Finally, how best to manage the storage and moving of materials within the shipyard.

Materials management has the main objective of ensuring that all the materials and parts for the construction of a ship are received and will meet the requirements of the owner and regulators which place quality and operational requirements on these items. For successful construction, the shipyard management needs the items to be made available to the production departments of the shipyard when required. Within the constraints of the market for the products required and regulations, there is another shipyard objective which is to minimise the cost of what is purchased.

As the shipbuilding industry has developed, so has the supply chain. Particularly over the last fifty years, the proportion of bought-in materials and services has been increasing. More complexity of the ships produced is one of the causes of increased purchasing from sources external to the shipyard. Attempts to save costs have also led managements to outsource many activities which they previously carried out in the shipyard. All this has to be managed alongside reduced production cycle times. Purchasing is therefore increasingly important for the ship construction industry and each individual shipyard.

The proportion of shipbuilding costs represented by bought in materials and services is now typically around 60% for most commercial ships, and up to 80% for some specialist types, of the total cost of the product although this does vary. Historically, because shipyards were the largest producers of steel structures and often the only organisations requiring large propulsion machinery and specialised pumps and other equipment, most shipyards produced their own requirements. So a

G. Bruce, *Shipbuilding Management*,
https://doi.org/10.1007/978-981-15-8975-1_8

typical shipyard would be close to iron ore and coal and have an engineering division alongside the ship construction. Woodwork, sheet metal and what used to be limited electrical outfitting facilities were also to be found on a shipyard site.

Purchasing has to interact closely with other activities both in the shipyard and outside, all carried out by stakeholders in a ship construction project. These interactions determine how effectively the purchasing is carried out. In managing external sources, the overall objective is to provide the shipyard with a supply of bought in items, at an acceptable, preferably low, cost. These must be delivered so as to be available to production when required and clearly to the correct specification. There are also many regulatory requirements to be satisfied by the equipment and materials. Reliability of supply can sometimes be as important as price.

Internally the main issues are the organisation of purchasing, how it is integrated with other functions and how well it meets the requirements for effective production. Considering the internal factors first, many past problems with purchasing have been caused by these. That is the internal operations of the shipbuilding company have an impact on the ability to carry out effective purchasing.

Creating detailed technical specifications based on past supplies of the equipment and products which are required may limit the choice of suppliers. It is essential for the technical department in a shipyard to maintain contact with suppliers and general trends in the market so that potential product improvements are not neglected. The specification may also cause quality or delivery problems. A specification which is difficult for a supplier to achieve may result in increased cost because of a need to perform rework. An onerous specification may also be responsible for late delivery, and the quality of the product which is finally delivered may not be as specified. An extreme example can be for military ships, where a military specification may result in a very high price because the supplier has to make items specially, whereas often a commercially available product would be sufficient.

The planning of the shipbuilding project may be faulty. In some cases the senior management may ask for a delivery which is too early for the supplier to manage. This can be for several reasons. There may be delays in developing the design of the ship, so that information the supplier requires is not available in time. The shipyard may have offered an unfeasible delivery date for the ship in order to secure the contract, making all the planning unrealistic. Shipyards are often seeking short delivery times for products because they are under pressure to deliver a ship in a short time period.

There are times when commercial imperatives may have to overrule what is feasible to provide. Given the opportunity many shipyard activity managers will ask for more time than is absolutely essential to prepare information. What is important is that the real situation is known so a genuinely informed decision can be made. Realistic purchasing lead times must be used for planning of ship projects.

Turning to external factors, a supplier may offer a delivery date which is hard to achieve and a shipyard may accept this date. Suppliers may be overloaded with work, but reluctant to refuse an order. Again, the parties are trying to operate to unrealistic schedules. In some cases there may be late changes which will affect the delivery date.

Minimising costs is a priority for all companies, and trying to drive a hard bargain with suppliers is often favoured as a strategy to achieve this low cost. The problem can occur when the focus of the company is simply on obtaining supplies for the minimum price. If the shipyard tries to drive a hard bargain, and the ability to do so will be highly dependent on the market conditions, then the supplier may seek loopholes in the specification to reduce the quality. A low price may be offered and then the supplier will try to find every ambiguity in the specification to claim an extra payment. If nothing else this will sour a relationship and distract shipyard staff from more important issues of materials supply.

Late payments to suppliers may cause future problems. Large companies often try to impose very onerous payment terms on small suppliers. If this is the case then supplier co-operation in future may be difficult to find.

Lack of discussion may limit the level of supplier service. A supplier may have potential to assist a shipyard by offering alternative products, revised designs or simply assistance with installation and commissioning.

The purchasing requirements are built into work breakdown and scheduling. The production schedule determines when delivery of materials and equipment is required. The purchasing lead time determines when an order must be placed and the technical information is required in time to place an order. The organisation of these functions, must take these factors into account. The work breakdown for production must take account of delivery times, in that early outfitting cannot take place if the necessary items are not available.

Supplier selection has been of increasing importance as the supply chain has grown. As the scale of shipbuilding increased, some of the manufacturing was moved to specialist suppliers. These suppliers were able to take advantage of economies of scale in their production. Particularly since the 1950s in most cases the scale of external supply to the shipyards rather than internal manufacturing increased significantly.

There are still shipyards which as part of large conglomerate companies are closely linked to suppliers, in Asia especially. For most shipyards however the selection of suppliers and managing them has become increasingly important. The relationship between a supplier and a shipyard can have an effect on the ability to produce ships of adequate quality, economically and on time. There are three basic relationships possible between a shipyard and a supplier. This is a simplified model of the real life situation which can have more complications.

In the first case, the shipyard is relatively large, and its business is very important to the supplier. In the marine business, the relative size and power of the supplier and customer may vary. There are some very large shipyards where the shipyard dominates the suppliers and can dictate terms and conditions of supply. Many suppliers are relatively small, local in their market and reliant on a single, or relatively few shipyard customers. Often there are several suppliers available to the shipyard, all competing for the business. In this case, the shipyard dominates the relationship and can largely dictate the terms of business. Small sub-contractors and suppliers of easily obtained items are typical. The relationship is one where the power lies with the shipyard. As a result, the shipyard is able to bargain for low prices and the supplier has little or no power in the relationship.

At the same time, small shipyards in Europe or elsewhere may have a very different supplier relationship. There are often "clusters" of small and medium sized companies which provide a service to shipowners in a region or port. One or more may operate a dry dock, with the others supplying labour, services either direct to owners or through a main contractor. The management of the complete supply chain, often in the form of a "virtual" company, is increasing in importance and a marine company may be more of a project manager and coordinator than a supplier of engineering services. For example there are publicly owned docks around the World which provide dry-docking services. These clusters of shipyards may be able to bargain with large suppliers to obtain better terms and conditions for purchased items, because they require more.

In the second case the shipyard and supplier are both relatively large and are effectively inter-dependent. The supply of cabins and other internal structure for large passenger ships is an example. The shipyard relies on a supplier for the specialist outfitting and the supplier similarly relies on the shipyard for its business. The relationship is one of more equality, and the power is more equal.

In the third case the supplier is large, usually international in its market, and the shipyard is relatively small. The shipyard needs the supplier to be able to fulfil its contracts, but has no real ability to seek improved terms of business. The power lies with the supplier.

For many equipment items, the supplier is likely to obtain the business from whichever shipyard is able to secure the contract. Large marine suppliers, such as engine builders may be in a more powerful position than the shipyards, because they know an engine will be required, and they will be sure of the order from whichever of several competing shipyards wins the ship contract. This is even more the case where the shipowner wishes to specify particular suppliers for reasons of standardisation within a fleet.

A variation of this relationship is where the shipyard, or indeed several shipyards, is a relatively minor customer for a large supplier. An example is electrical cables, where the ideal customer for a manufacturer will require hundreds, even thousands, of kilometres of a specific cable specification and size. Then the manufacturer can set up production and manufacture long production runs very economically. In contrast ships require short runs of specific sizes and a variety of specifications. Even where a larger shipyard or perhaps a group of shipyards require cables, the total quantities are still small. As a result the shipyard has no power to influence the prices asked.

Underpinning all the discussion is the question of price. This always has importance, because the fundamental objective if the shipyard is keeping the costs of production as low as possible. Historically, price was seen by shipyards as a major consideration. Where there are multiple suppliers for any given product, selection on price is sensible, given equality of products. However there has been considerable consolidation of the companies which supply the shipbuilding industry and other considerations are required.

It is important to re-state that obtaining a low or at least reasonable price remains important as a means to control the costs of ship production. Given the different relationships which can exist, how they can be managed to achieve the shipyard objectives is an important question.

Quality of supply was in the past largely measured by inspection at delivery. The need to produce ships of high quality, to meet regulations, the needs of a shipowner and for ships to be built in a short time, have placed much more emphasis on ensuring the supplied items will be of good quality. It is essential to avoid rework or at least keep any that arises to a minimum.

Delivery reliability is another important consideration. A low price and good quality do not assist if an item which is critical to the timely completion of a ship is delivered late. Here it is important to take into account supplier relationships. That is, using a supplier regularly and building a close understanding of their operations and problems can result in a better service. Where several shipyards are able to use co-operative purchasing to place larger volumes of business with a supplier, the reliability can be improved.

First looking at price considerations, the objective for a shipyard remains to acquire the materials needed at a low price. This low price can be achieved by several strategies, including hard negotiation, wherever the power in the relationship lies with the shipbuilder. Also the shipbuilder can threaten or actually change suppliers, in a competitive supplier market. Finally there is the potential in some cases to seek the lowest specification which will allow the ship to achieve regulatory approval.

Where the power relationship is more equal, or even if that is not the case, an alternative approach to obtaining lower prices can be achieved. Other strategies include building strong relationships with suppliers. This requires continuing to use particular suppliers, often jointly forecasting production levels. This will allow the suppliers to plan their production more effectively, with the objective for the shipyard of allowing them to offer lower prices. This does require the shipyard to have an adequate level of orders placed with a supplier, or to be part of a group of shipyards using the same supplier.

There can be joint reviews of equipment specifications to seek mutually beneficial changes which reduce costs while maintaining functionality and acceptability to owner and regulators. Shipyards can also agree long-term contracts with one or a few suppliers. The second approach may offer the best long term results, for both parties, but is much harder in terms of the effort required. It does rely on each of the two parties seeing they will gain benefits from a long term collaborative relationship.

Quality assurance has an important role to play in material supply. Inspection on delivery can identify whether an item, or a batch of items, is acceptable for use in the shipyard production. If the batch is not acceptable then it can be rejected. However any consequential losses to the shipyard cannot be easily managed. Items of comparatively small value can seriously affect the completion and quality of the completed ship.

Again, building relationships with suppliers is one strategy to use to ensure that quality is managed and unacceptable deliveries do not occur. If a good relationship is established, there can be detailed discussion of requirements and specifications to ensure that the supplier has the best opportunity to meet the shipyard requirements at acceptable cost. Joint inspection of items prior to delivery is important, especially for major items such as main machinery which will be factory tested prior to delivery. Other items can benefit from such inspections.

Where the supplier and customer are large and volumes of supply are also large and critical to the ship completion, joint measures to improve production methods can be adopted. This has been more common in the motor car industry where small numbers of suppliers produce large volumes of products.

Equally, again dependent on the relationship between supplier and customer, there is a need to the shipyard to carry out supplier evaluation to review how well the supplier performs. Issues including delivery reliability and quality are considered. Where feasible a change of supplier can be implemented if quality is not adequate and then fails to improve.

Delivery on time matters because late delivery of equipment or materials is often a serious problem for a shipyard trying to adhere to a production schedule. The consequences often go beyond the immediate item concerned because in the worst cases, the whole project may be delayed. While the cost of a particular supply can be recovered by the shipyard, this may be a very small part of the total ship, and the shipyard may incur a penalty from the owner.

Steps which can be taken to reduce the potential for late delivery include accurate and realistic scheduling, by both shipyard and supplier. Asking the impossible of a supplier in terms of delivery date is rarely successful. Equally, suppliers have been known to make delivery promises which are not achievable.

Avoiding engineering changes is important, as these can seriously disrupt a supplier's production schedules. The supplier should also avoid making product changes without prior warning to and approval from the shipyard. Holding joint progress reviews is effective, and visiting a supplier premises to view progress can improve a relationship, lead to useful discussion and sometimes lead to a means of faster delivery.

The shipyard must still carry out initial supplier appraisal, conduct on-site inspections for important equipment items and if the power is available and alternative suppliers are in the market, make changes.

In some cases the shipowner may want to provide some items of equipment to the shipyard. This is a potential source of problems, starting with the shipyard needing to know precisely what will be delivered. The shipowner is not likely to have a large technical staff, so those there are will probably be overworked. They may assume that the equipment supplier will understand what information the shipyard requires in advance of delivery. If some additional items are needed for installation, then the shipyard has to know that as well. Communication with the supplier may be only via the owner, so the shipyard may not be able to discuss directly with the supplier, and even if this is allowed, will not be able order anything additional. If a foundation is needed, who supplies this and all the other interfaces between the ship systems and the equipment? There is more to the potential scope of supply than the owner may realise and all extra items will be a potential for disagreements. This can extend to necessary tests and trials, especially when the equipment is integrated into the ship.

The use of equipment supply by the owner carries risks, and if any risks occur, then there may be an impact on the schedule which has been previously agreed by owner and shipyard. The owner's site representative is not likely to accept that a delay is their fault, so further disagreement is likely.

Overall, having the owner provide equipment is not a good idea, though there is another side to the question. This is where the ship requires some highly specialised equipment with which the owner is familiar and the shipyard is not. Then a risk appears that if the shipyard places the order, they may seek a low price and not fully understand the requirements. In these circumstances, it is probably sensible to create a separate contract for the specialist installation.

In summary, materials supply is of increasing importance as industry conditions change. Success is dependant on integrating procurement with design, planning and other functions and effective organisation using information technology. The technical functions must recognise the need to release information early in some cases to allow the equipment to be procured in time for production. Discussion with a supplier can determine exactly what information is needed for them to create an order, start preparation and then move into production. The supplier can be requested to provide information such as the footprint of equipment supports and connection locations early to allow the ship design to progress. Joint efforts can simplify the issue.

Then good suppliers should be identified through formal appraisal procedures. These need not be very complex but will guide the shipyard, subject to their relative power, in selection. Creating partnerships, formal or informal, with suppliers can help to secure price, delivery and quality. Whatever is done, it is essential that the shipyard continues the monitoring of levels of service.

The material management role continues once the supplies are received in the shipyard. Inventory is the material and equipment, purchased for contracts, which has not yet been used in production. Inventory costs money, so that minimising inventory is a useful objective. There are however still good reasons for keeping some inventory, some of the most important of which are listed here.

Inventory first ensures items of equipment and materials are available to production when needed. Having a small stock available in the shipyard is useful and gives time to make quality inspections where necessary, to collate items into packages according to the needs of the work packages, and to make these available to the work stations in good time. This inventory will also provide insurance against late delivery from suppliers.

Having an inventory also allows purchasing in economic batch sizes. If a larger number of items can be ordered at one time, then there may be potential for a discount on the price. The number of orders and that cost can also be reduced. There will then be a need to store the items which arrive early, so there may be costs of extra storage and materials handling to partly offset the savings in price.

Once production is in progress, inventory can be created between successive activities. Often, succeeding activities operate at different speeds and also need to process items in a specific order to ensure efficient production. Inventory then allows the activities to operate with independence from each other. On the other hand, a large inventory of work in progress is often a sign of problems such as poor planning.

In some cases the rate of production of assemblies and other items for the ship is constant, because this allows for a stable workforce and efficient production. However the rate at which the items are required on the ship under construction is often variable. For example most of the steel units or blocks are required at the early stages of ship construction, to provide several working areas and also to have areas of the ship available for outfitting as early as possible.

In this case it is necessary to assemble most of the ship units steadily, and then store them so they are available when required. This is a case where some inventory is essential. Advantage can be taken of the units being available in storage to install outfitting before the units go to the construction dock. For a large shipyard with several ships under construction simultaneously this is less of a problem, but for most medium and small sized companies the inventory is required.

Inventory is also used to permit some flexibility in local scheduling. If the local supervisor can produce items of similar size and type as a batch, even if some are not required for some time, then efficiency can be improved.

There are costs associated with inventory, which must be taken into account when deciding the correct inventory levels. First there are holding costs which include the cost of providing a suitable building or other facility to keep materials and bought in items safely and in good condition. Facilities may not be fully used for the whole time, or it may be necessary to rent facilities from time to time.

Then there is an associated cost of materials handling, since items my need to be moved during storage. In the case of steel plates, if there are large numbers in storage then it is usually necessary to sort the plates in order to select those which are required for immediate production. Materials in storage will require insurance, which may be high in the case of valuable items. Some items may become obsolete after a prolonged period of storage. There may be a danger of breakages for more fragile items in storage.

The costs of inventory have to be balanced against material shortage costs, which are the costs of lost or delayed production because an item or items are not available when required. In general it is safer to have a larger than a smaller inventory, especially for items which may affect the critical path for the ship project.

Where there is limited inventory, there may be extra production set-up costs. These are the costs of changing production from one to another product. A typical shipbuilding example is the production of pipes. Where these require to be bent, the pipe bending machine has to be set up for a specific diameter and specific bend radius. Changing over to a new setting is time-consuming so once a set up is ready it is desirable to complete as many pipes as possible. If this is not done then the time to make changes will increase and production will be lost. The other side of this is the need to store a lot of finished pipes for some time.

Management of the appropriate inventory level is a question of balance between two often conflicting requirements. These are to ensure that production is not delayed by shortage of materials and at the same time to try to keep the inventory cost to a minimum.

The key to successful management of inventory is as with many aspects of ship construction the planning of work. For large items of equipment, and materials which are bought to order the planning system determines when they are required for production. To achieve the required availability, the items must be delivered to the shipyard earlier. This requires orders to be placed with suppliers in good time, so planning also determines when technical information must be available. The planning process also determines the necessary order date. However, there will almost always be cases where the process is not straightforward, and where these are not identified as project risks and then reduced, there will require to be management action to deal with exceptional requirements.

For the inventory management, various models are available to determine how often orders should be placed and in what quantities. The two basic models are fixed quantity ordering and fixed time ordering.

In general a minimum level of inventory is set for the items, which will ensure that there is never a shortage. The level is set according to previous experience of the suppliers' reliability, whether the suppliers are close or not, possibly overseas and to manage possible variations in demand within the shipyard. Then a maximum level is set which is designed to balance two issues. The cost of re-ordering items is increased if more frequent orders are placed. The cost of maintaining inventory is increased if fewer and larger orders are placed with suppliers, because that leads to high numbers of items in storage.

There are alternatives depending whether future demand is certain or uncertain. If it is certain then fixed time ordering can be used, so that an order is placed at regular intervals. If the level of demand is uncertain then fixed quantity ordering is used, generally the difference between maximum and minimum inventory levels.

Once the material has been received in the shipyard, it has to be delivered to the production location. The production is split into small work packages, each with a bill of materials, along with other work station information which has been developed at the detailed stage of design and planning. Work study will have determined the sequence of work, the planning system will have identified when the materials are required and the technical departments will have prepared the material information for purchasing purposes.

Once a work package is authorised for production, the materials required are identified. They should normally be in storage, as the order placement and delivery date will have been pre-planned. If any shortages are identified, or if special items are identified at a late stage then these are expedited. Normally the production authorisation is made three to four weeks ahead of the planned production date to give time for the expediting if it is required.

As part of the management of materials, how items are managed within the shipyard is of considerable importance. Various estimates of the cost of materials handling have been made, but none appear to be based on specific studies. Including local manipulation of parts and assemblies during the assembly process, estimates vary from 25% of labour costs up to 50% in some cases. A UK government study identified estimates for industry in general of between 20% and 35% of total production costs in the USA and Europe. The definition of materials handling varies but it is considered prudent here to take in all movements.

Included in materials handling therefore are all necessary, and importantly unnecessary, movements of materials after their arrival in the shipyard. That is from receipt of all piece parts and raw materials from storage through processing to output storage, of all interim products and of units or blocks, part or whole ships and ship sections.

The term materials handing includes transport, storage, sorting, location and manipulation of any item in the shipyard. How materials handling is carried out depends to a large extent on interim product characteristics. Product characteristics which will have an effect on materials handling are the mass, dimensions and geometry. Mass determines the capacity of the handling device, which can be anything from a human to a vehicle to a large tonnage crane. The dimensions dictate the size of vehicle or pallet for small items, and also whether a crane used to move the items has two hooks to make load handling safer. The geometry has an effect on the balance of a load.

The density of the material being carried also affects the equipment to be used, with small, dense loads such as castings on pallets. The material is also a factor, including robustness so a steel item is less susceptible to damage. Steel and other magnetic materials can be picked up by magnets which make the handling of large steel plates much easier.

The value of items is also considered, mainly for their storage requirements. Sadly items of high value do sometimes go missing, and so a secure store is essential for them. Careful packing and handling are also important to avoid damage. Materials handling equipment has to be selected to suit the requirements of the industry. There are numerous means for routine transport and handling available in ship production, many of which are found in a range of industries, including those discussed below.

Cranes have been the usual means of moving materials in shipyards for many decades having evolved from simple derricks. They have the great merit of flexibility in use, because they can be used for all elements of handing, basically moving loads, stacking loads and positioning loads accurately. A number of fittings are available for cranes, primarily hooks, magnets, spreader beams for long or large loads and specialised fittings such as fork lifts for pallets.

Conveyors are useful for moving large numbers of items along fixed paths, but they can interfere with free movement of people and other transport. A common use is to transfer steel materials from an unloading area to storage and then on to preparation. In a large shipyard the distances can be long and the conveyor saves investment in cranes with lengthy travel distances. However it is not a good solution for most handling issues, unless there is some overwhelming benefit.

There are alternatives to conventional fixed conveyors which can serve the same purpose but avoid blocking other transport routes. There are intermittent conveyors where a gap in the conveyor line is bridged by a wheeled platform with conveyor rollers on top. It is also possible to use automated guided vehicles, which follow a guide wire in the ground. However for a large number of movements over a long distance they may not provide sufficient capacity because of the need to return empty for the next load in a sequence.

Rail-mounted vehicles are used where there is a frequent flow of materials along a fixed path. They often are a good alternative to conveyors for distributing materials to different work areas. Examples are for moving treated and painted steel plates from those processes to various cutting work stations. The benefit is they do not obstruct roadways for other vehicles and people. Automated guided vehicles are used in many industries to move items from stores and around various work stations. They do not need any drivers, which is a cost saving. Specialised versions can be used in stores for order picking. They are not common in shipyards, except for very high levels of production.

Vehicle use in shipyards has evolved from simple hand carts or horse drawn carts via the use of railways. Vehicles have the general benefit of great flexibility in operation, being able to go in any location where there is a road or other paved surface, for example in a workshop. The cost of roadways is of course part of the total cost of providing the transport.

Many conventional vehicles are used, such as flat bed trucks and enclosed vans to deliver materials from storage to work stations around the shipyard. Within workshops, fork lift trucks are common and give an ability to lift and position items as well as move them. Fork lifts are ubiquitous in industry generally, and standard models are used in shipyards. There are also some more specialised vehicles, and ancillary equipment, which are required for the often large loads and odd geometries found in a shipyard. Side-loading fork lift trucks have an elongated load platform and are used for long loads, such as steel profiles and pipes.

There are also various types of auxiliary equipment, which enhance the cranes and vehicles. These include pallets, to carry several items at once, crane attachments, including lifting forks so the crane can safely handle pallets.

Efficient production is dependent on materials flow. The basis of any manufacturing process is a set of work stations, linked by a material flow system. The more complex is the process, the more important is material flow. As ship production has developed, with more stages of production, the correct provision of materials to each stage when required has become fundamental. There are some basic principles to be applied in developing materials flow. The list here is not exhaustive, but includes the main ones.

First it is necessary to plan all handling activities. This may appear obvious, but regular observation in many shipyards will demonstrate how frequently materials are picked, moved and placed without achieving any useful outcome. If each movement is planned, then first only necessary movements will be made. Secondly, lifting an item provides an opportunity when it is put down to take it to a work station, sort it into a particular production sequence, reorient it ready for the next production activity or accumulate several items to create a unit load.

To incorporate necessary reorientation or sorting operations with movements can give large savings, basically by reducing the number of handling operations needed to take an item to its next processing activity and present it in the best orientation.

Where large numbers of items need to be moved between activities, then the distances between them should be as short as possible. If a long movement is needed because of the size of a shipyard site, then careful cost analysis of the most appropriate transport method is required.

The transport should be mechanised where possible, which is really most of the time. Apart from health and safety considerations, using human labour to move items is expensive and results in those engaged on the work not being able to carry out a more productive activity.

Minimising the idle time for handling equipment is very important since the equipment is relatively expensive to acquire and maintain. Its purpose is to move materials and time not so doing is wasting the investment. This includes pallets and other equipment which is often used for storage rather than movement. Regular observation can reveal how little some equipment is actually used. Just as planning work station activity is important to productivity, planning transport increases productivity.

Some additional actions to make the best use of vehicles, cranes and handling aids can be taken. The first is to make use of "empty" return journeys, where a transport means which has completed a movement is used for something else. It could be removing waste material from a work station. The use of unit loads will maximise the efficiency of transport, by reducing the number of movements required for small items.

A means of saving of initial investment in transport is to minimise the equipment to payload weight ratio. A vehicle generally weighs between ten and twenty percent of the payload it can carry. A crane on the other hand will be between five and ten times the weight of the payload. As the weight of the transport medium is a good proxy for the cost, it is better to use a vehicle where possible. A crane used for a simple load movement is not using its full capability, which includes lifting and manipulating a load.

Some shipyards have managed to combine the two to minimise investment. They have a fixed position crane which minimises the cost by avoiding the tracks, traction motors, wheels and power feed. Then vehicles are used to move materials and assemblies as required, including to the crane for example to turnover a large unit, or to assemble a block from units. A single large capacity crane can be used, arranged so that it can serve several work stations which would not usually be co-located.

Developing an efficient materials flow is important. An overall shipyard layout is developed as part of the shipyard strategy, whether this is a new, green field development or a re-development of an existing operation. The layout is initially based on estimated movements. Once the layout is established and effort moves on to specific shipbuilding projects, further analysis is needed.

The first step is to identify all items which will flow through a facility. The first items are materials, which should include scrap and waste, as this has to be cleared from work areas and prepared for disposal. It is important to also consider movement of equipment, especially when this is required aboard ships during final outfitting, when unplanned movements can be time consuming. People and information also need to move.

Then it is necessary to quantify the data for all the movements. As ship designs and production planning develop, information becomes available for the movements required.

Review any additional factors which need to be taken into account. Safety should always be considered so that the planned materials flow does not compromise this. There may be a need for flexibility, if more than one ship is under construction, or any variation to the programme of work is likely. It may also be necessary to consider future expansion of the facilities and this may dictate for example use of vehicles rather than conveyors.

Then use an analytical technique. There are many of these which can be useful, divided into two basic categories, qualitative and quantitative. The techniques are part of work study, which is discussed in Chap. 14 as part of assembly activities.

Work sampling is often very effective for transport studies. In one study the company reported that fork lift truck transport was a problem, with long waits for the vehicles when required to move finished items between processes. The site was old and larger as a result than would have been designed from new, so the activities were often distant from each other. The solution proposed was to hire more vehicles to improve the position. However a work sampling exercise was carried out. Over two weeks the status of any fork lift truck observed randomly was noted, either stopped, empty, stopped with a load, or moving with a load. Whether a stopped vehicle had a driver present was also recorded.

At the end of two weeks the data was reviewed and it was found that the vehicles were more likely to be stopped than in operation, despite the belief that the delays in transport were due to a shortage of vehicles. This led to further investigation which found that the vehicles were largely kept in individual workshops to ensure their instant availability to that workshop.

The solution was to remove the vehicles to a "cab rank" central to the site where they could wait for a call, then return on completion of the movement. The final result was fewer, rather than more fork lift trucks, with a higher activity rate, and importantly very few if any delays. Fuel consumption increased a little but overall a large cost saving was made.

The next stage is to develop alternative flow patterns for the movement of materials. The total number of movements along alternative flow paths, the distances moved and other data can be calculated. Then the alternatives can be evaluated using any objective criteria which are useful. The total distances moved or the number of items for the distances can be used as examples. It is also necessary to take into account any constraints on movements. Site constraints may exist, slopes for example which restrict larger vehicles. If conveyors are to be considered then they may obstruct the potential paths for other handling, especially vehicles.

Using the evaluation, then select the preferred alternative, which may include elements from different attempts. Finally review the arrangement again against the desired criteria. Then draw the proposed flow on a plan of the site. It is important to consult widely within the organisation, and perhaps use external expertise as well in order to identify any possible bottlenecks or problems.

As an extension to work study, once an apparently satisfactory handling scheme has been developed, the use of discrete event simulation is an excellent means of checking again for bottlenecks and also fine tuning the arrangements.

Chapter 9
Ship Design for Production

Engineering is the art of doing for ten shillings what any fool can do for twenty!
—Wellington

This book is primarily about the management of the ship construction process, so the design of a ship is for another book. However, some consideration of the design process is important, in that it can have a significant effect on the efficiency of the production activities. Preliminary design information is the basis for deciding the processes and methods to be used, planning the work, selecting facilities or where necessary sub-contractors, identifying and acquiring the necessary resources.

The production of a ship depends on the generation of a large set of information about that ship. Historically, technical departments were concerned primarily with vehicle function, and production information was developed within the production departments. The steel production departments had to be skilled in interpreting drawings the main purpose of which was to demonstrate the structural capability of the steelwork. All the information needed to actually make the steel parts was developed in the mould loft and workshops. Similarly the arrangement drawings showed the location of equipment and some pipe work, but most of the information required to make and install pipes for the ship systems was created in the workshop.

The primary functions of the design process include the development of a hull form for efficient propulsion, adequate sea-keeping characteristics and stability. Further requirements are a structural design which meets strength requirements while keeping weight low and compliance with all regulatory requirements. The design also includes all the engineering systems to ensure the above, including propulsion machinery and supporting auxiliary equipment, cargo handling equipment, whether liquid, solid, bulk or packaged. Accommodation, navigation and any specialist equipment, for example process plant are also developed. The significance of the different design elements depends on the ship type.

So the design activity is initially concerned with the definition of the end product, and making certain that the product will conform to the specification and meet all necessary regulations. To ensure that the construction of the ship can be carried out

G. Bruce, *Shipbuilding Management*,
https://doi.org/10.1007/978-981-15-8975-1_9

efficiently, design needs to go beyond this and include consideration of the production requirements.

Design is carried out in several stages, for convenience here five stages are indentified. The first three are intended to fulfil the traditional design functions to create a functionally suitable design. The final two are intended to ensure the design is correct for production, that is it will be as easy to construct as possible consistent with working correctly.

Although for convenience the stages of design are presented as sequential, in fact production is now an important factor in the information developed at each stage of design. In the activity map in Chap. 3, the five stages can be identified in the design process. Different shipbuilders may use different names for the stages, so the important consideration is the definition of each one. Here the stages are,

Conceptual design, when the features of the ship which meet mission or owner requirements are developed. The conceptual design may be prepared by a shipowner or operator according to their needs, by a shipyard seeking to develop a saleable design or by a third party specialist designer contracted by either. The output of the stage is a design which is worth taking further because it meets initial operational and cost criteria. That is the ship is expected to produce a profitable outcome for the builder and the future owner.

In preliminary design the features of the proposed ship are developed to the stage at which they are sufficient to form a contract between an owner and a shipbuilder. Again, the design may be developed by an owner, a builder or a third party. The level of detail can vary, depending on the precise relationship between the parties to the potential contract.

At the functional design stage, the features of the ship are developed so as to be able to gain owner, regulator and classification approval. This represents the end point for the design process as traditionally viewed. It is usually the end point if the design is undertaken by or on behalf of the future owner. However, since the overall objective is to build the ship once it has been designed two further stages are required.

Transitional design is a stage which was identified formally and named in the 1970s, based on US studies of shipbuilding in Japan. The transition is to convert the design features of the ship from a system orientation to a product orientation. By system is meant a functional element of the ship, for example the ballast water system or the structural system. The transition objective is to view the ship, or a ship system, not as an operating entity but as something to be produced. During the production process, parts of the ship, often including parts of several ship systems, form what are called "interim products". Where the whole ship is the final product, an interim product is created at an earlier stage of the production process. An interim product may be inverted for ease of work access so design information will require particular consideration. It often contains elements of several ship systems. The functions of these for the finished ship are less important here than the specific needs of production.

Although transition design is defined here as the stage at which production is specifically considered, it is an important consideration at all stages. Since production is primarily about combining parts and small structures into larger ship elements (interim products), how easy or difficult this is to carry out is an important factor in the speed and cost of production. Early consideration of this, for example at the concept, preliminary and functional design stages can help reduce costs considerably.

At a very early design stage, the geometry and dimensions of units, blocks and other structures can be determined, as can the routes, preferably straight, for pipes and other systems. These features can be refined as the design develops through subsequent stages.

The objective of detailed design, the final stage is to define the features of planning units sufficiently to support production. The later stages of design are primarily to develop production information, from the sequence of work in assembling parts to the precise dimensions of a part, for example a pipe.

The outputs of transitional and then detailed design can be summarised. First there are working drawings, which define each interim product, from a small part to an assembly to a large ship unit. Fitting and assembly instructions are provided where required, as are all quality checks and measurements to be made. Each interim product needs a material list to ensure all the materials needed for that product are made available. The materials and production equipment are specified so that production is carried out correctly. All material requisitions are needed so that externally sourced parts can be acquired.

Overall the design process creates information which defines what is to be produced which, in addition to the ship and its systems, includes the purchasing of materials and equipment and part manufacturing.

Efficient production needs a close match between the production activities and the interim products. This leads to the concept of design for production, which has the primary objective of reducing production costs to a minimum, consistent with the ship fulfilling operational requirements. It is important to note that there is no basic conflict between different design criteria. Design for operation and design for production can be compatible. In principle, a good design can also take into account design for maintenance and even, potentially, design for disposal.

Design for production is necessary primarily because of the need to keep the cost of producing a ship to the minimum. The reduction of costs is achieved largely by meeting two goals. The first of these is the reduction of inherent work content in the design. The second is to ensure that the ship design will make the best use of the available production facilities.

It is a responsibility of the design function, but depends on close co-operation between the design and production departments of a company so that there is adequate data from the production function. Creating multi-disciplined teams with both design and production staff is one way to ensure the cooperation. Where the ship design is within the supply of the shipbuilder, this should be easy to achieve. Where there is a third party design, this becomes more difficult because design to make a ship easier to produce does require extra design effort and therefore costs. These costs are offset by reduced production costs, but when there are more parties to the building project it becomes more difficult to apportion the costs and benefits. A sub-contracted design house will not benefit from any work to simplify production, and the owner is unlikely to pay for such work.

Design for production must both support and respond to changes in production technology intended to reduce the costs or increase the output of production. For an existing shipyard, with no plans to change the production facilities, the design has to fit the existing processes. So such aspects as maximum material size, crane capacity, welding processes and building dimensions can act as constraints on the ability to vary a design. Where there is going to be a significant change of the ship products, and some investment is planned, then that investment can be aligned with the design to achieve best production outcomes.

In other cases a completely new, or substantially upgraded shipyard, may be planned and this will provide an excellent opportunity to align design and production in the production analysis. Examples of supporting efficient production include making the assembly sizes and arrangements suitable to make maximum use of specialised production systems. Examples which can be incorporated include a flat steel panel line, a web line and curved unit assembly jigs.

One of the key features of design for production is the use of standards which allows more repetitive processes giving opportunities for continuous improvement.

For steelwork, a number of actions can be carried out in terms of standardisation, First, standard material sizes and properties can be adopted. This will simplify specification of the material to be purchased and may present opportunities to the reduce costs of that material to the shipyard. All steel plates may be of one size, or just a few lengths and widths. Arguably there may be more waste material, but in most cases the cost of labour is a more significant item than the cost of steel. Also off-cuts of steel are often useable during the production process. More controversially, steel plate thicknesses can be limited, for example using only 2 mm increments. This would require careful consideration at the design stage, but is potentially possible.

Standard sizes can reduce the potential for shortages, in that any piece of material can be used, rather than a specific piece for a purpose. The costs of storage can be reduced, because there is less need to sort material by size and so less storage area may be used. The need to sort material into a specific order for production also carries a cost and this can be at least partly eliminated by using standard items.

Maximising the sizes of purchased material can also reduce work content. Simply put, larger pieces require less joining, so the welding workload can be reduced. This may have considerable impact on the facilities need, for example the crane lifting capacities, but should be an important consideration in any development plan for a shipyard. In a high wage area, equipment can usually replace labour to reduce production costs. Where wages are low, the cost of equipment may be harder to justify. The speed of ship construction may also be a factor.

Once steel material enters production, there can be standard preparation requirements. The design can incorporate standards for piece parts such as brackets, or for cut-outs in stiffeners. Standard process technology can be used, which can lead to increased automation. Where pipe runs are designed at an early stage to be straight as far as possible, there is an opportunity to use standard pipes. Overall the use of standards can be expected to improve a number of aspects of the production process, all of which would reduce the costs.

Process flow can be simplified, with fewer variations in the sequence of processes need to complete a part or assembly. There will be reduced need for sorting parts, as where they are standard any part can be selected and used. The parts will be interchangeable, so if errors occur and a part has to be rejected or reworked, another part can be substituted. Overall it can be expected that there will be improvements in performance and quality. As a final consideration, training requirements may be reduced, and moving back in the overall process, there will be fewer individual parts to design and standard items may be reusable for different ships.

At the sub assembly stage several examples can be considered. The use of standard components can be built into the design, for example for brackets between side frames and tank top. Standardised assemblies may be more suited to assembly using jigs and fixtures to make the work more efficient. The standard assembly size can be more compatible with materials handling equipment.

At the unit assembly stage consideration moves from the standardisation of items to the improvement of the assembly processes. The first example is to design the processes to give self-supporting units. These simply can be assembled and once secured in place will require no external support (from cranes or props for example) while the various sub-assemblies are welded together.

A second example is to design the units and the assembly process so there is a minimum need for staging and access equipment. This will generally require that the unit is rotated during the assembly process so that the work of positioning, securing and welding the elements of the unit together is always at or close to floor level.

In addition to the steel structure, the units can be designed so that their completion includes outfitting and coating, which will also reduce the need for access after the unitsare built into the ship. It is possible to design the supports for ship systems and equipment into the structure, or to locate and weld them in place during the assembly process, again keeping as much work as possible at floor level.

Even at the ship construction stage, simply building inherently similar designs will show a benefit, as the workforce gains familiarity with the production requirements. This uses the learning curve effect, where repeating a process results in more efficient operations, and therefore faster, more accurate production. Standard ships, or ships which are very similar in design, become progressively easier to build.

More significantly, the decisions taken at an early design stage can influence production costs. By considering the product work breakdown structure based on the manufacturing strategy from the start of design, the impact on production costs can be minimised. The impact of early production consideration will influence how the structure can be split into blocks, units and assemblies. This can have beneficial effects.

It can reduce the number of blocks which are taken to the building dock. This will usually reduce the total time taken for construction in the dock. It can simplify the connections between the blocks, which reduces the welding required, improves access to welds and makes the process selection easier. Blocks are usually coated before construction, with the exception of areas which would be affected by welding, so reduced welding will also reduce the final coating requirements. Access is then only required for joining blocks or units and the same access system then serves the coating process.

Finally, the design can maximise the weight of blocks within crane limitations. Larger blocks will make outfitting easier and more complete.

There are some pre-requisites for design for production. If it is to be effective, beyond small scale adoption of such items as standardised brackets, a number of features are needed. There must be a recognisable manufacturing strategy for the shipyard, incorporating both standard processes and stable processes.

There must be a feedback system from production to design. This can identify problems caused by design features, for example where connections between parts are difficult because of a need for unnecessary accuracy. The feedback can also identify opportunities to improve design features and ensure that information for production is presented in a suitable format. Because the designers may not be at the right stage of work on the next ship, their immediate interest in a novel design may be limited. The feedback system should store information until it is needed. Overall the objective is to provide non-confrontational feedback so that there is benefit to the production processes.

There must be a ship definition strategy which offers specific outputs at each stage of design. These outputs are specific to the requirements of the stage of work that has been reached and the needs of the particular user of the information. In particular, production information in a suitable format is required to simplify work for the production workers.

There must be a production engineering function which manages facilities and process development, based on the feedback results. The production engineers also have a part to play in developing accurate estimating of labour requirements. This will then lead to improved scheduling of the work to be done.

Design for production has specific objectives. Overall the intention is to make the work of ship construction faster and lower cost. The two key objectives are to reduce the work content in construction of a ship and to reduce the cycle time to construct a ship. Secondary objectives, also in support of speed and cost reduction, include improved interim product quality because of the easier fitting and thus reduced rework and an improved working environment with better access to work pieces.

All the above should also improve material utilisation by reducing any waste where processes are not carried out correctly first time.

The process requires the design function to work with production from the earliest stage. The designer still is responsible for the development of the ship as a whole, or a collection of systems such as hull structure, cargo handling and propulsion. The ship must always comply with owner and other requirements.

It is necessary to consider the ship in terms of work breakdown structures. Several of these can be identified. The designer begins with a system work breakdown structure which considers the functions of the ship and each system. For materials purchasing, a purchasing work breakdown structure is used. Here the focus is on effective purchasing, without concern about what the function of the purchased items will be. For production, a product-oriented work breakdown structure considers elements of the ship during the production process and is intended to secure efficient production.

So far this discussion has focussed on how the designer can assist production operations. To achieve the cost reductions, production information is needed which is accurate, up-to-date and in a form which the designer can use readily. If there is not a close interaction between designer and producer, with good communication on both sides, then design for production will remain an aspiration.

Information on facilities which is required includes the capabilities of individual work station, the characteristics of the shipyard cranes such as capacities, lift heights and outreach, and the capacity of multiple crane lifts for very large units and blocks. Building characteristics are also important, for example door sizes and other transport restrictions. Information is also needed on all the processes in use, for example welding capabilities and pipe bending diameters. All this is also part of the feedback system.

Outfitting also requires early consideration of production requirements. The key is to identify potential outfit assemblies (modules) from the earliest stage of design. The identification of service routes is also important, because if this is done early, then pipe runs can be made largely straight. This simplifies manufacture of the pipes and when they are installed improves access to each system. Access for future maintenance can also be part of the process. This is rarely done perhaps because the benefits are not appreciated by the owner and no value is seen by the shipyard. This is mistaken.

Further benefits are that common foundations for equipment and services can be developed. In addition to running pipes and other services run in parallel, work can be moved from installation on ship to a workshop environment with less travelling for workers, reduced material handling and better access for installation. Testing of parts of systems can also be carried out before on-board installation, so problems can be found and fixed at a more convenient stage of production. Also, interference between systems can be largely eliminated.

The accuracy and repeatability of the production of assemblies is also important. The preferred production methods, and therefore the preferred design characteristics, are influenced by accuracy. Stable and accurate production allows complex three-dimensional blocks to be produced, with the benefits which were outlined earlier. Further, large scale installation of outfitting on blocks can be completed. Overall this allows for the minimum of work remaining at the final construction stage.

On the other hand, poor accuracy of assembly does not allow for large blocks and early outfitting, because there will be necessary adjustment when structures are joined. If there are large, rigid assemblies then such adjustments become very difficult. Poor accuracy will lead to simple panel units, with welding left unfinished so that the panel alignment can be adjusted. Smaller units, such as panels reduce how much outfitting can be achieved. As a result there will be large quantities of work to be completed only at final construction, where efficiency is much lower.

Production engineering is a shipyard function which also has a major input into cost estimating. At the build strategy stage of the contract there is preliminary design information available. At this stage a firm price is needed so that a contract can be considered. As mentioned earlier, the level of design development on contract signing can vary. For the purposes of this discussion, it will be assumed that the design is sufficiently advanced for a specification which will include the major equipment and machinery items. The arrangements of the ship will be sufficiently advanced to allow the lengths of pipe and other systems to be determined with reasonable accuracy. Also the areas and volumes of spaces in the ship will be available so that estimates of quantities of materials can be estimated.

At the tactical planning stage, overall production information is available and this allows more detailed estimates to be made. Accurate resource requirements must be decided so that the project can be planned and executed, but this will be within the budget already set by the contract price. There may be agreed variability in the costs where the project retains some uncertainty.

As the design develops an accurate quantity can be determined and at the detailed design stage there is a precise definition of each steel part to be used. It must be remembered that the steel cost is invoiced steel, not used steel. There may be 10% wastage from part cutting and this must be added to the net steel identified from accurate CAD information.

Particularly in the preparation and assembly stages of a ship, there are large numbers of interim products to be made so the learning curve theory does apply to many shipyard activities. Improvements are as a result of learning, and not making mistakes, but it is also important to analyse the activities and seek improvements in the processes used.

The role of production engineering is very much to review the production activities and the processes within them to seek improvements. The review of a process will begin with a measurement of the productivity, using what ever measure is appropriate. This will establish a starting point, and the measure added to all the others in the shipyard will become a basis for future detailed cost estimating. As has been noted, to remain competitive a shipyard needs to improve productivity, more or less continuously.

So considering the process model for an activity, it may be that the output in terms of quantity and resource requirements, especially labour, is not meeting requirements. At this point production engineering is required to make a systematic and stable improvement. For this purpose, the techniques for layout development and work study are all potentially relevant.

Where a process can be improved, discussing this with the labour force engaged in it is a good starting point. They use the equipment and carry out the work on a daily basis, and their understanding of it will be superior to any gained from a short term study. So it is necessary to engage the workers in the process so that their understanding can be combined with professional analysis and knowledge of potential new equipment.

At this point it is necessary to consider the motivation for management. Ideally, this is to improve the productivity by reorganising the process. Assuming this is to increase production, then worker co-operation should be easily available, because increased production has at least the potential to increase wages. Also an improved process is likely to have better equipment, for example manipulators to present work pieces in a convenient orientation. This has the benefit of improving the working conditions, reducing worker fatigue and reducing potential dangers so should be beneficial to everyone engaged.

If the wish is to reduce employment, or obtain more production at lower cost, then engaging with the workforce becomes more difficult. This book is not in any way seeking to judge the behaviour, simply to present views on how the development of a process can take place and what should be considered. However, to consider this in project terms, and re-developing a process is a project, the workers are stakeholders whose objectives need to be taken into account.

The development of an activity therefore, will have several stages. The first is to measure current output and productivity as a benchmark and then decide on the required output and productivity for the future. The second is to analyse the current processes using a suitable work study technique and determine the elements of the activity which can be revised. This may be sufficient to make an adequate improvement. This is often the case so it is very important that this analysis is made before seeking more costly investment.

If the improvement is insufficient, then the third step is to look for alternative equipment, either production process or materials handling, which can offer the production increase. Fourth, a revised arrangement can be created, incorporating the alternative equipment. Further work study can then determine whether this will meet the production objectives. The fifth step is to make a financial analysis of the revised arrangement and consider whether the payback on the necessary investment is sufficient. It can be that the equipment is actually too expensive to be justified.

If the revision is financially sound, then the change can proceed. Once the new activity is in place, it has to be monitored to ensure that the results are meeting expectations.

In one case a shipyard was considering the purchase of a production line for flat, steel panels. This included conveyors to move the panels from work station to work station, fairing systems and single side welding equipment. In seeking a financial justification for the investment, the initial focus was on welding processes, because the equipment seller offered a faster process. Before proceeding the company carried out some work study on the existing non-mechanised activities.

The analysis showed that the current welding process was more than fast enough for the length of the welds to be carried out, but the welding being achieved was only around 30% of what should have been possible. Then by improving the materials handling of parts onto the work stations, the alignment and the fairing of those parts, the current process could be better used and the planned investment was unnecessary.

A summary of design for production is that it seeks to reduce the cost of constructing a ship. The means of achieving this reduction is to reduce the work content of the construction by simplifying the interim products and by standardising the details. This is one part of the overall cost reduction effort. Figure 9.1 shows how the work content reduction works with two other aspects.

The reduced work content is carried out using the most efficient processes, so that less work is required and it is done well, with high productivity. The other aspect is the effective management of the business, so that the best use is made of the

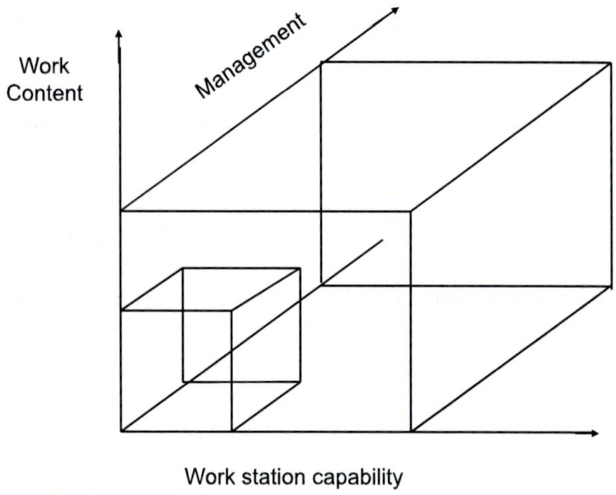

Fig. 9.1 Reducing work content

resources available. Figure 9.1 shows three aspects of reducing production costs. The vertical axis represents increasing work content due to poor design. Lowering the work content is a first objective of design for production. The next axis represents the capability of the work station, where reducing the hours required will be the objective. The third axis considers the management of the process, basically reducing time lost because of lack of materials or information, idle time and any other lost production.

So in Fig. 9.1, the minimum work is done, using the best processes and with those processes used for a high proportion of the time available. The effectiveness of each is multiplied by the others, so the higher cost can be represented by the volume of the larger of the two cubes. Reducing the work content, inefficiencies and lost time is represented by the smaller cube.

Chapter 10
Planning

A goal without a plan is just a wish
—Larry Elder

Planning is a fundamental component of the overall shipbuilding management process. It allows the management of a shipyard to decide in advance when a project will be completed. This is initially to confirm that the shipbuilding project can achieve the delivery date specified, then to be sure that all the component parts of the project from start to completion will also be achievable. Progress monitoring, described later in Chap. 17, then compares the actual progress against the plan, and is a basis for decisions to amend the production process if delays occur.

This chapter offers the reader an overview of shipbuilding planning and how it is organised in shipyards. Some of the potential problems are identified and how they may be overcome is explained. There is also a very brief overview of some important planning tools. These are bar charts, as the simplest form of planning, logic diagrams to show the sequence of work on a project, Gantt Charts for the scheduling of work stations and two network planning techniques, which are Programme Evaluation and Review Technique (PERT) and Critical Path Method (CPM).

Many shipbuilding projects are large and their life from initial enquiry to completion of production is spread over several years. The planning of these projects is very diverse and covers a wide range of activities. The planning is generally carried out in stages, which correspond to the development of a project. In essence, the detail of the planning increases as time passes and the detailed definition of the project becomes increasingly available. For the purposes of this book, four stages of planning are identified, one for the overall shipbuilding company and three for each individual project.

Corporate planning generally provides an overview of the planned future for the company, looking around five years ahead as part of the company overall strategy. It is useful here to be reminded of the variations in ship types, sizes and timescales for construction. A small shipyard may have a corporate planning horizon of only perhaps two years as the timescale of each project is generally shorter. A large shipyard may

G. Bruce, *Shipbuilding Management*,
https://doi.org/10.1007/978-981-15-8975-1_10

look further ahead than five years. Examples are large cruise ship builders and naval shipyards where the forward horizon may be ten or more years.

The corporate plan will set dates for current projects and projected dates for future projects It will also plan the use of major resources, for example dry docks. This provides a framework for the projects to be undertaken and also identifies constraints on those projects. The corporate shipbuilding plan is an important part of the corporate strategy, developed on a rolling basis to help identify future strategies and any constraints that will need to be overcome.

Within the project start and completion dates set by the corporate plan, and the constraints identified, the project planning can begin. Key dates are identified within the overall timescale set for the project, including when can the design activities commence, as previous project designs near completion. This helps to determine when materials, major equipment and fittings can be specified for purchasing. The availability of materials determines when the preparation and assembly work can begin in production. Availability of assemblies determines when work in the dry dock work can begin, and leads to the date when the ship be able to or will be required to leave the dock. Other key dates are when the ship should be substantially completed for trials, which precede the required delivery date.

Strategic planning covers the duration of an individual ship project. It usually looks ahead from the start to the end of a project, typically a period of two years. Using the key dates above, and the available design and work breakdown information available as the preliminary design for the ship is developed, the dates of the main supporting activities can be determined. This uses past performance data and looks more closely at the availability of facilities. The output will be a set of planning events which are significant points in the project. Examples are a key date, such as first unit to the dry dock, a substantial piece of outfitting, such as the installation of main generators or the completion of a space, for example a cargo hold or accommodation area.

The final output from strategic planning is a key planning event network for the project. The scale of this is project dependent, but taking a typical mid-sized cargo ship, the network will have around 100 planning events. The planning will also determine the overall resource loadings, both facilities and people. Resources are assessed from estimates made previously. The use of a standard sub-network for each planning unit allows the timing of resources to be determined. The sub-network is a logic diagram showing the sequence of activities and their completions, leading to a planning event.

These will usually not vary from ship to ship, so for example for a steel unit, events are the completion of the necessary design information, the availability of material for the unit, preparation and cutting of the steel, assembly stages, testing, outfitting, coating and completion before finally delivery to storage ready for the dry dock.

The strategic planning output has to be aligned with other projects to ensure there are no clashes of requirements for the facilities and other resources. As a result the plan may have several iterations before an acceptable version is reached. There is always a danger of a company accepting a project for commercial reasons, without sufficient scrutiny of the resource and other constraints. This will create risks to the successful completion of the work on time and within budget.

Tactical planning is the next stage of planning the project, setting the timetables for the completion of the work leading up to the planning events. The strategic plan has set the demand for when work for a project must be completed in the various departments of the shipyard, and by subcontractors. The sub-networks provide due dates for the interim products. Tactical planning then schedules the work on these interim products, typically covering the next few months in each department and may also look at part of a project. It typically looks ahead three months though this is a variable time, again depending on the scale of the project and size of ship.

The complication in tactical planning is that it has to include several projects, at least in a larger shipyard. That is because typically one or possibly more projects will be reaching the end of work in the department, one or more will be in progress, and one or more will be about to start work. If any project is running late, or requires some re-work or extra work, then there is an immediate conflict with the subsequent project or projects. If a late project takes resources from the start of a new one, then the completion of that project in turn is threatened.

Individual project managers will necessarily focus on their own work and demand priority, so it is essential that any problems are managed by the department to complete as much work as possible on time. If there is a conflict between projects which cannot be resolved, then it must be referred to a higher authority for a solution.

The manager of the department may also see his efficiency as the most important outcome. In some shipyards, this may be the benchmark by which the departmental manager is judged, for example how many tonnes of steel or what number of pipes have been produced in a given timescale This is likely to be in conflict with the project management objectives, because efficiency is most easily achieved by carrying out all similar work at one time whereas the project managers will see the timing of each production item on time as the priority. The problem is discussed later when the scheduling of work is addressed.

Detailed planning considers individual work stations or work groups. It is carried out often weekly, to create a rolling programme of work typically for four weeks. There are a wide range of work stations so at this level planning the four week period is again considered typical, but not absolute. This time horizon is to allow time for checking that materials and other resources will be available when required for the work station. For example a check will be made that materials have been delivered and are correctly scheduled within the time period covered.

At this level, the tasks to be carried out and the necessary order for this should have been decided, so the local manager, or often a foreman or supervisor will have as his objective the completion of work to the specified time, and of course within the man-hour budget and to the quality requirements. It is sometimes the case that items have been missed from the task list, or additional tasks have been identified. These are usually classed as urgent, at least by the project manager who identifies the need, and so this is one way in which the smooth, planned operation of a work station can be disrupted. A recurring need to insert unexpected tasks indicates a failure elsewhere in the shipyard, perhaps late design work and should be investigated.

A measure to mitigate the effects is to set up extra work stations to deal with these unscheduled requirements, which are generally small scale, leaving the main planned work undisturbed.

It is worth repeating that other timescales may be used for different product types and more or fewer levels of planning may be found, depending on circumstances.

The organisation of planning will often vary from company to company. In some cases, all planning is in a centralised department. This is perhaps a result of comprehensive planning software which is increasingly available, where huge amounts of data can be aggregated and manipulated in real time. However, centralisation is not usually the best arrangement. In some cases it is popular with senior management because it seems to offer them a means of checking up on project and department heads. So a central approach may be due to a level of mistrust of the local management, or simply a desire to feel or to be seen to be in control of a project.

The software in a centralised planning setup can easily generate revised schedules for all areas, and this can be destabilising for the local work stations. It is more or less impossible to build all the constraints which exist locally into a central schedule. As a result a project managed this way is likely to go increasingly out of control. Micro management by senior executives is rarely successful.

Although centralisation allows a project manager to review the status of any, or all, tasks on a continuous basis, the problem also arises of what it is possible to do with the information generated. Essentially the manager's capacity to read, understand and use information on the status must equate to the quantity of information to be managed. If planning is decentralised then much of the decision making can be spread among a larger group of individuals, giving the greater management capacity. Good project management assumes there are well trained managers and supervisors, who can be trusted and in turn trust their seniors.

It is thus generally a mistake to try to centralise planning at the more detailed levels, because of the volume of information to be processed and the speed with which local conditions may change. A centralised planning department is likely to spend most of the time trying to keep up with small, local changes rather than actually planning. So generally planning at the detailed level, and often at the tactical level, is localised. Local knowledge of the department or work station should allow the planning to be effective, provided the local management is correctly trained. This is a very important aspect, so that the local supervisor is confident in scheduling work and then managing the outcomes. It should also give the higher management confidence that the local ability exists, and perhaps most importantly, that any problems will be reported quickly to a higher level for resolution.

In many companies in the past there has been a reluctance to report problems, because of unsympathetic senior management. The effect has been to hide problems while local supervisors and the project management attempt to find a local solution. This is always difficult because they are not able to see a bigger picture and so will not understand the impact of the initial problem and the actions they choose to take, on the overall project. Their actions may also affect several other projects.

Realistically, good training, good support to newly appointed supervisors and acceptance that problems may not be caused by individuals should result in improved outcomes. Mistakes can provide a useful learning experience in a supportive environment.

For a department, a specialist planner or a small planning group can be made available, reporting to the management of the department. This planner will use the software which is available, and its plans and schedules will effectively be integrated with the overall plans. However the planner should have flexibility to organise local schedules within some general constraints. Localised planning does depend on the local management reporting any significant deviations from plan to the next higher level of management.

Strategic planning is generally centralised, but may report to a project manager in a large organisation. There may then be a corporate planning level which co-ordinates several projects. A comprehensive planning system is useful in co-ordinating the various levels of planning.

Bar charts, Gantt charts, logic networks, CPM and PERT networks are common tools used in project planning. Product breakdown structure and work breakdown structure define how the large project is split into manageable elements to be used for planning and control purposes.

The bar chart shows the activities for a project or part project against a timescale. It provides an instant visual appreciation of the work to be done, and with current and required dates highlighted, is easy to use in identifying progress. It is limited in what is shown but highly effective to display progress and lateness clearly.

A Gantt chart is a popular project management bar chart that tracks tasks across time. First developed in 1917, as part of the "scientific management" movement, the Gantt chart is able to show both time and interdependencies between tasks, and this is now its normal use. Since their first introduction, Gantt charts have become a standard project management tool, because they are easy to use and can display a lot of information. As such they are used to show the phases, tasks, milestones and resources needed as part of a project.

Arguably the Gantt chart is most useful when it is used to show the use being made of resources in the shipyard, whether the people in work groups, physical facilities such as workshops, docks or individual machines. For each resource the scheduled tasks are displayed as bars alongside the timescale. This is useful in allowing the management to readily identify underused or overstretched resources, which demand action.

A logic diagram indicates the sequence of activities in a project over its duration. In the first instance it is not time dependent, it simply shows which activities have to be completed before others, and hence the sequencing of the work to be undertaken. It shows which activity logically precedes or follows another activity. For one-off projects, the logic diagram is very important.

For industries such as ship construction, there may be alternative logical sequences, driven by different circumstances. However a logic diagram is still the starting point for a project network. The logic diagram is very good for developing sub networks for the interim products which have to be made prior to achieving a planning event on the main network.

When the logic diagram is complete and the project management is in agreement about the sequences shown, it can be converted into a network by the addition of times for the tasks which have been identified. It can be used to identify the milestones and critical path for a project which are a result of the dependencies in a project. The overall timescale for the project can also be calculated.

Critical path method (CPM) was developed by Dupont for the construction of chemical works. The series of tasks which make up a project can be represented on a network, where CPM assumes that the time for each task is predictable accurately. This is the case where there is sufficient past data from experience to predict the times required. The method is basically a logic diagram with timings for the activities.

The program evaluation and review technique (PERT) was developed by the United States navy for the management of the Polaris ballistic missile submarine project. PERT is a method for analysing the tasks involved in completing a given project, especially the time needed to complete each task and identifying the minimum time required to complete the total project. Given the risks and uncertainties in such a challenging project as Polaris, putting a nuclear reactor into a submarine and equipping it with missiles, timescales could not be certain. It is the management of variable timescales for activities that is the fundamental difference between CPM and PERT.

A danger in having probabilistic timescales in a plan is that as the project is started and then progresses, the times tend to drift towards the longest estimate. On the other hand, the method does recognise the potential for delays. Any uncertainty in the timescale must be reflected in the strategy for the project and importantly should be manageable using the contract.

The basic steps in a network analysis can be listed. First it is essential to identify all the activities required by the project, which is based on a suitable work breakdown structure. Next the precedence relationships are identified, showing which activities must be completed before others can begin. This gives a logic diagram. Then the expected time is determined for each activity, which may be an accurate estimate or a probability. These first steps require a lot of thought and effort to ensure the results are as good as possible. The analysis which follows is relatively straightforward, using planning software. The danger is in believing the computer output if the logic or timings are faulty, or worse if any activities have been missed.

The remaining steps are to develop a network diagram, then identify the earliest start and finish dates for the activities, followed by the latest start and finish dates. The critical path can then be found which determines the total time to complete the project. The process should be iterative, so that an early, simpler version of the network is available to ensure that the proposed contract dates for a ship project are feasible.

A work breakdown structure is a hierarchical decomposition of the deliverables needed to complete a project. It breaks the deliverables down into manageable work packages that can be scheduled, costed and can have people or other resources assigned to them. A work breakdown structure is a standard project management tool and the basis for project planning.

In project management, a product work breakdown structure (PWBS) is a hierarchical structure of the components that make up a project deliverable. For ship construction, the PWBS is most important in shaping and controlling the production processes. It is described here in some detail.

A work breakdown structure can be described as a convenient way to divide a large complex project into manageable elements. This is important to successful project management. A fundamental principle of the WBS is that it is based on deliverables, which are usually described as tangible, usable and complete. Examples of deliverables for a ship project are the same the events previously described, including completion of a design package, an outfitted block, ready to lift on to the ship, a tested ship system, or an outfit module.

The WBS is arranged as a hierarchy of deliverables. The highest level is the complete ship, ready for the customer. Intermediate levels include a ship system, structural block, major module or information package. Examples of deliverables at lower levels of the structure are a hull unit which is one of several which will be joined to create a block, a sub-assembly which will be part of one of the hull units or an individual piece part which will be part of an assembly.

In the context of a ship, the deliverable is often described as an "interim product", which is something which is complete as defined, can be supplied to a customer, internal or external.

The WBS hierarchy is not necessarily a single structure. It can be split into levels, in various ways according to the needs of the project at a particular stage. During design the breakdown is generally into ship systems. For assembly, the breakdown is generally by hull structure but also includes some outfit assemblies. During outfitting installation, the breakdown may be done by geographical zone of the ship, or by functional zone.

Each part of the ship belongs to more than one breakdown structure, each of which has some attributes of the part. For example a pipe piece may have several places in the overall breakdown, according to the stage of activity on the project. During design, the pipe is part of a system, where the attributes are for example, pipe wall thickness, diameter and material. The pipe will then become part of a material package for purchasing, where the attributes include supplier, specification, cost and quality. This package may be for more than one project to allow purchasing at a low cost. In the shipyard, the pipe is stored, then delivered as part of a work package with which it is fabricated, where production attributes matter, for example diameter and bending requirements.

In assembly, the pipe is then part of a module to which it is attached, where the important attributes are location and structural attachments. This module may have further assembly, as part of a hull unit in which it is installed, where the important attributes are again location, attachments and specifically alignment of adjacent pipes. On board the ship, the pipe is part of a zone within which it is tested, where the specification of the pipe is again important. Finally, the pipe reverts to the design system for operational testing of the ship.

The work breakdown for any project must be tailored to the specific needs of that project. There is no universal "correct" answer to the form of a work breakdown. The

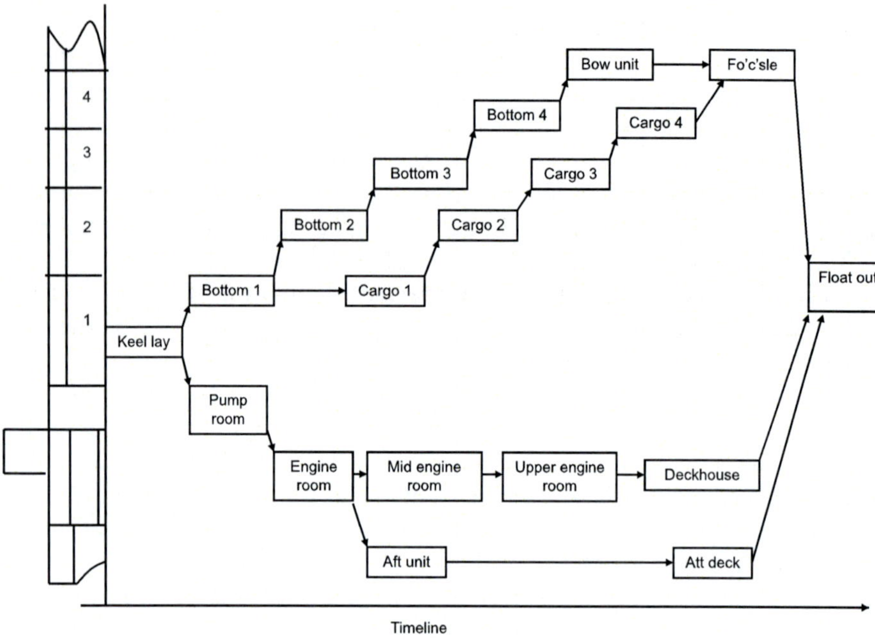

Fig. 10.1 Strategic network

work breakdown for a project has an effect on several aspects of a project. Project management is product oriented, and deals with delivery on time, delivery within budget and delivery to specification to complete a project. A deliverable-based WBS gives form and structure to project planning. At the strategic planning level, the deliverables forming the project network must link to the deliverable milestone at the corporate level.

The WBS develops in stages as information becomes available. The initial breakdown makes broad decisions, dividing the ship into the planning events for the network. The timing of equipment installations and of system completions is also included. The basis for the strategic network is the programme for the construction site, which is driven by key contract dates in the corporate plan.

Figure 10.1 shows an outline network, where the events are the installation of each structural unit or block. The format shows the logic diagram, which is often created by the production engineering function from analysis of the most effective ship construction sequence. This logic then has timescales applied, showing when each unit or block can be installed on the ship. The times depend on how soon after a unit is installed it is aligned and secured in its location. For most shipyards the times will be based on past projects and initial analysis and estimates. The times will be fixed, so as to give a clear definition of when the delivery and all preceding events will occur. It may be possible to use variable times for a prototype ship, but any uncertainty in delivery will have to be in agreement with the owner and form part of the contract.

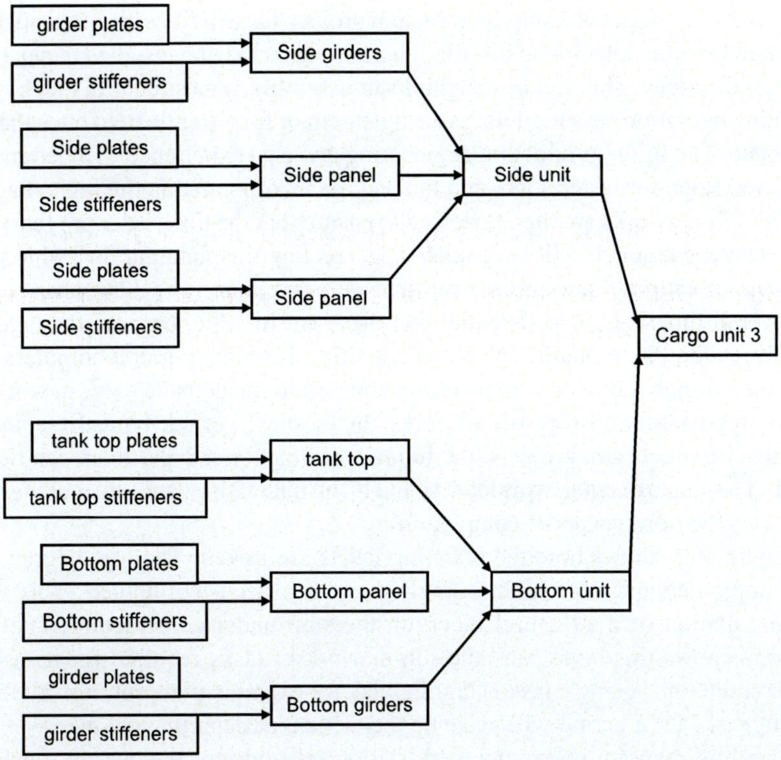

Fig. 10.2 Sub-network logic diagram

From the above the overall time that will be taken for construction is found, and the timeline for construction is shown at the bottom of the figure. This will provide a useful visual understanding of progress as the construction happens. There is also a schematic of the ship which also helps with monitoring progress.

Having started with the actual ship construction, this network can then be developed to include the timing of major design, procurement, information and test and trials events. Typically then such a network has up to one hundred planning units which finish with the required contract delivery. Any additional planning units are identified for major elements of each functional activity, or major sub-contracts or external services.

It is always better to keep the network to a reasonably manageable number of events so that the logic can be clearly seen and is certain. However, more detail will be required, so for each planning unit, a more detailed WBS is developed. The activities required to complete the planning unit are identified, usually with each representing a stage of work for the planning unit. Figure 10.2 shows a standard logic diagram, and when lead times based on past projects are added, a sub-network is created. This is similar to the main network, and so that the planning and then progress will be driven by dates, the events on the sub-network are the completion of each stage.

At the early stages of a ship construction project, the activities are often not defined in detail. For example, for an installation zone, items may be installed at any one of a number of stages. This can be on a hull sub-assembly, on a structural block, prior to a turning operation, as a module, as a single item or for a fragile item once the ship is complete. The initial production engineering and past experience will determine the preferred stage for installation, and this will be incorporated in the logic diagram.

The sub-networks are then reviewed to ensure that the timescales are feasible and the resources required will be available. In creating the planning for a ship project, the contract estimate determines overall resources required and the contract dates set the overall timescale. It is essential that these are feasible, because they are major constraints on the remaining planning activities, for planning unit completions and then the sub-networks for each planning unit determine demand dates in workshops.

By aggregating all the sub-networks for the next period, typically some three months, for all current projects, the future workload on a department can be calculated. The departmental workload is likely in many shipyards to include several projects at various stages of completion.

Figure 10.3 shows how the planning levels are linked. The top diagram shows a strategic network on which a single event has been highlighted. This may be the installation of a structural block on the ship under construction. The second diagram shows this block, and the sub-network of tasks required for its assembly and completion. A single task is highlighted, for example plate cutting. All the plate cutting tasks for a period will make up the tactical or departmental plan.

The third diagram shows the work station schedule for the cutting. Each of the cutting tasks is shown on a bar chart. The tactical plan has not yet been adjusted so the tasks may require to be completed at the same time. Also the schedule will need to be revised to obtain the best outcome.

The next requirement is to schedule the work for each department and then each work station in those departments The problem for the departments is to meet the demand set by higher planning levels. The overall loading which has been determined during the estimating process only represents the mean load on the facilities. When the demand schedules are produced, based on due dates, this may result in uneven demand. As a result, day to day scheduling is essential to even out the loading on the work stations as far as possible.

In general there are alternative strategies to overcome any remaining resource overloads. For much of general industry a steady flow of products is made, and the precise timing is not important because the products go into stocks in a supply chain. So for many the possibilities include re-scheduling the work to an earlier or later time. This is difficult in shipbuilding, because to create schedules at an early stage of a project assumes the design and hence production information will already be available, which is not usually the case. Later scheduling of production may cause delays to a shipbuilding programme. Any rescheduling of work also assumes there is sufficient capacity in a work station if the timing of tasks is altered.

Overtime working is possible if the overload is short term, or sub-contracting can be used if a longer term overload is found. This assumes that the work is suitable for sub-contracting, that there is availability when required, that the quality and the

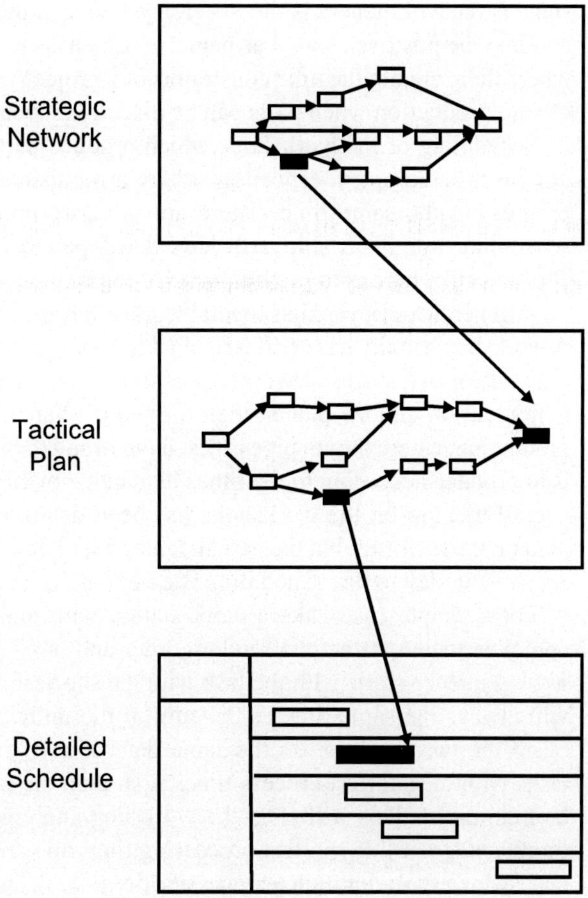

Fig. 10.3 Planning and scheduling links

delivery will be acceptable. It is possible to increase capacity if the extra work is expected for the longer term but there will be an increased cost of investment.

Before committing to any investment, sub-contract or overtime cost, it is important to achieve the best schedule possible with the existing production capacity. This scheduling includes the sequencing of the work and assigning start and finish times to tasks. There are several criteria which can be used to judge schedules. The first is the makespan, which is the time taken to process a given set of tasks, so minimizing the makespan is a possible objective. The flow time is the time any one task spends in the workshop, and minimizing this may be a useful criterion to provide work for a subsequent work station. Both of these are concerned with making the workshop efficient, by processing tasks quickly.

Ship production scheduling also has to account for completing work on time, as many of the components to be made are unique. Tardiness is the length of time by which the completion date for a task exceeds the planned due date. Note that if a task is completed early, the tardiness is zero. Early completion is generally of no particular benefit, because the output can only be used when the next production stage is ready. Lateness is the difference between the completion date and due date and may be positive as well as negative. Lateness is a useful measure for industries where there are not the time constraints of a project, and can apply to the early stages of ship production when parts can be placed in inventory for a short time.

Scheduling of the workshops which produce parts, small assemblies and larger units can be complex, especially where a number of different ship projects are in progress at the same time. There are software products which will assist in the scheduling, and many shipyards have developed their own systems and procedures. The objective here is to present some of the basic problems and solutions.

In general, a shipyard has several workshops, each with a number of work stations. A workshop usually has dedicated machines or work stations and there are frequently only one or two stages of work. An example is plate preparation, where a steel plate is first cut on a flame planer, then formed to shape in a shell rolls. Another is pipe production, where the cut pipe is first flanged and then bent to shape. The requirement is to produce according to the times demanded by higher planning levels, where the overall loading on the workshops has been determined. This represents the mean load on the facilities, but the actual demand schedules may result in uneven demand. As a result, day to day scheduling is essential.

For a simple case, take a work station with one machine and several tasks to complete which arrive in a random order and have different processing times. The work sequence starts with the task with the shortest processing time. The makespan will always the same time as the sum of the individual process times, so delivery of all the tasks will be on the same date. If a rule applies to always process the tasks with the shortest process times first, then the average flowtime for a task will be minimized. This will clear the workshop and complete the first task quickly. A disadvantage of this shortest processing time rule are that in a busy workshop, new tasks with a short duration may get priority, so tasks with long process times are keep waiting. Importantly, no account is taken of due dates, although if tasks do leave the work station more quickly, the next process will have work available.

If the tasks above also have due dates then if they are processed in order of their arrival, there are likely to be tasks which are late and others may be early. If they are processed by shortest process time, some tasks will be very late. However the key measure is tardiness, where late tasks create delays but early tasks offer no benefit. If the tasks are processed in an order with the earliest due date first, there is likely to be a longer average flowtime, and more lateness, but the average tardiness should be reduced which is usually a good result.

The above are very simple examples, whereas the typical shipyard has flow workshops, with several machines or workstations, generally with tasks moving through them in a standard sequence. Assuming the processing times on machines can still vary from task to task, the makespan will again always be the same for the first machine, but if earliest processing times on that are long then the later machines may have idle time. As a result the scheduling is more complex but the same basic rules can be applied. For example to minimise the overall makespan, the tasks with the shortest times on the first machine should be completed first so that the second machine can begin operations as soon as possible.

The objective generally is to achieve the due dates for tasks first, then to minimize the overall time taken and so improve efficiency. In real world workshops the priorities and conditions may vary if new tasks are acquired. The tasks may depend on the progress of others, and tasks have to be prioritised for several projects. The effect of the schedules on the overall project management has to be taken into account. While basic scheduling rules can produce a large improvement where the current process is not very effective, the complexity of the shipyard demands more and simulation is often the solution which is adopted.

Dynamic event simulation is a means of creating a life-history of the operation of a system and can usefully model completely situations which have some random effects. It has been available as a technique for many years, though there is now better accessibility through proprietary computer software products. These can now provide graphical models, with animation which helps to understand how a workshop can be expected to operate. A note of caution is that as with planning and other activities described in the book, using sound information and logic must still underpin the simulation process.

The elements of a typical simulation study are first to formulate the problem. This has to be done clearly, so that relevant data can be collected or estimated to define a model of the activities being studied. Data is based on the typical processing times observed in a workshop, or on work study for a new situation. This is analysed to calculate information such as processing times and how these vary due to operating problems. The frequency of materials and interim product arrivals at a workshop are also needed. It is essential to review the validity of this information with the operators who will best understand how the system works. The frequency of arrivals and the variations in processing times can be used to generate probabilities for use in the simulation model.

Once an understanding is gained, it is possible to build a computer model and carry out some pilot runs. These should replicate the estimates or past data which will confirm the validity of these initial runs and therefore of the model. Once the model is operating satisfactorily, the next stage is to design the experiments to be carried out. These will explore alternative schedules, arrangements of machines, processing speeds, numbers of workers and any other variables in the workshop. Then a number of test runs of the model are made as required. The results are analysed and the best solution selected. Finally the results are documented and put into operation.

There are advantages and disadvantages of simulation. The technique can provide an insight into how a system will behave in reality, including how random events can affect operations. The use can avoid possibly expensive mistakes in investment. An important use is to identify bottlenecks in production and to improve operating procedures.

However, the cost of data collection can be expensive, the level of detail must be appropriate and the logic must be correct. Some of the more sophisticated simulation software can be expensive to operate.

Chapter 11
Cost Estimating

If we choose we can live in a world of comforting illusion
—Noam Chomsky
It's easy to see, hard to foresee
—Benjamin Franklin

The cost of a ship or any other large, made-to-order product must be estimated, in the first place to determine whether the construction of that ship is a viable project. This is necessary for both the customer who will own or manage the ship, and the potential builder. An initial estimate of ship cost is needed when the ship is no more than an option for the owner's future trading plans or a potential product to be part of the shipyard strategy.

Estimating problems exist because a ship construction project is by definition the creation of something novel. They include the following. Many ships are prototypes so there is limited data from similar past experience on which an estimate can be based. The product information which is available from very early designs is limited, but it is still necessary to create an initial cost estimate. As a rule the availability of design data lags behind the need for it in cost estimating through the life of a project.

There are also exchange rate fluctuations over time so the value of an international project is likely to vary from the inception to completion. In additional future cost inflation is unknown, as is the future supply and demand for shipbuilding materials and equipment. These factors will affect the actual cost of construction.

There is a degree of risk in any estimate because of the above. In fact the final cost of a ship is only known after it is complete, and even then will depend on the accuracy of accounting assumptions and data recording.

As part of the future strategy for the shipyard, the future product ranges must be considered. Assuming this is not simply a continuation of current products, with some minor variations then some new initial designs will be created for discussion purposes.

At this point there is very limited design information on which an estimate can be based, but the development of at least an order of magnitude cost is required. There is a need to produce a credible price for potential customers so that the shipyard commercial function can decide whether the ship is viable as a future product. There may be discussions with potential customers and some price indication will be

G. Bruce, *Shipbuilding Management*,
https://doi.org/10.1007/978-981-15-8975-1_11

required before these can proceed. In some cases there is a close relationship between owner and shipyard, for example for large passenger ships, for military ships and for many classes of specialised, non-cargo ships. Both customer and contractor need initial cost estimates to decide whether to proceed with the project or not.

At this point where the ship is just an idea, the order of magnitude cost will be used to review concepts, from which the most promising can be selected for further development. The accuracy at this stage for a prototype ship is limited, and the final cost if the ship goes ahead is likely to be 20% more or less than the initial estimate. This is significant, but will be adequate for initial product selection and will be refined if the product is selected for further study. This estimate will also be used by the shipyard for evaluation if there may be investment in the production facilities.

An approximate early indication of cost can be developed from main characteristics of the ship. For example, it can be based on the cargo capacity of a ship. Given records of past contracts a formula can be developed which relates the cost of a ship to the capacity. If the proposed new product is similar to previous ships, perhaps with a development in size, or with some novel equipment, then this is a useful starting point for the cost.

So to take a very simple example, if the intention is to develop a larger ship for transport of LNG, and there is a history of smaller ships then the following gives a first indication. If an LNG ship of 160,000 m^3 is proposed, and the shipyard has a history of ships of 125,000 m^3 a formula can be developed. This will be using the ratio of cargo capacity to predict the ratio of cost. However this only works for a reasonably consistent data set of past projects, in the same or similar shipyards using the same or similar technology, with the same cost base. If there is more data available, for example if steel, machinery and insulation requirements are estimated then a more sophisticated formula can be developed. However the same caveats will apply and the sophistication of the formula will only be useful if there is sufficient similarity between the two ships and the same shipyard is to be engaged.

The potential cost of a ship or other project may be estimated using a variety of estimating parameters. At the initial design stage weight is usually the parameter chosen although capacity is an alternative. Various parameters can be and are used in different shipyards. In all cases the value of the estimating method depends on accurate data gathering and analysis from past projects. Alternatives used in different shipyards include deadweight (dwt), lightship, gross tonnes (gt) and compensated gross tonnes (cgt). The basic weights of the ship will be part of the owner's initial requirements and later of the contract, so are reliable. For large cargo ships the steel weight is also a potential measure to use. All of these figures are estimated at the early stage of a project and can be reasonably accurate.

Figure 11.1 shows how plotting ship cost against the parameter can provide a regression curve. Cost is a function of the parameter and a formula to link the two can be developed. The data can be plotted on a logarithmic scale to give a straight line, which makes the information easier to interpret. However, the formula which is developed to produce a cost is only accurate in terms of the parameter and data selected and how good the data is in the first place.

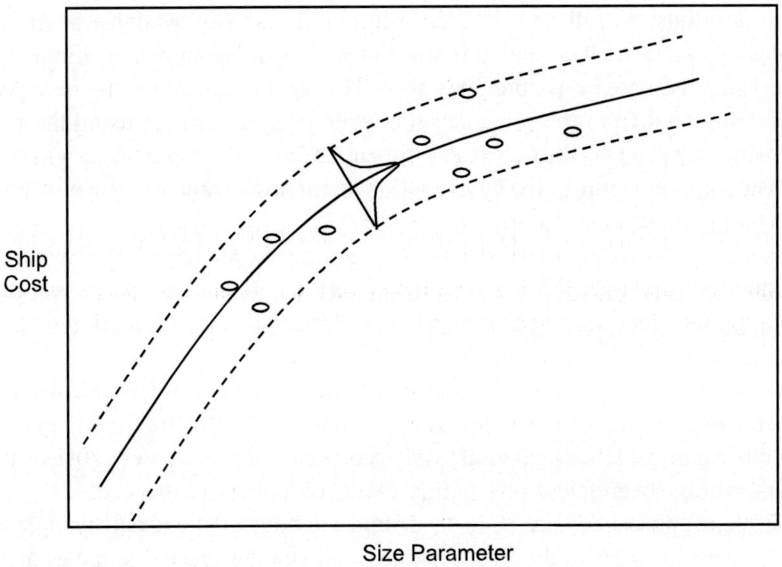

Fig. 11.1 Parametric estimating

There are dangers in using the parameters simply because there may be a high level of variability between the costs identified for the different past ships. The data collected generally produces a scatter of data points and the curve represents the best fit for these data points. There will be potential, if a new point on the curve is selected for a project, for the actual value to be significantly higher or lower. Ideally several parameters should be used to provide some capacity to cross check the estimates.

If an initial design is to be developed, because it seems to be of interest to potential customers then a design study will be carried out. This is where there is a potential project and the design can be taken to a stage where a contract may be signed. How far the design is to be developed is variable. In some cases the design if commissioned by the owner will be very complete and will be sent to prospective builders for a quotation. In others, there may be a joint development between a shipyard and regular customer, so the design development may continue after agreement to proceed. Typically at this stage of the design development, the estimated costs again for a prototype ship may vary from the final cost by plus or minus 10% to 15%.

The cost of a ship is made up of several factors. The first of these is material cost, which includes steel, machinery, equipment, pipes and coatings. Then there are sub-contracted items usually including electrical outfitting, heating, ventilation and air conditioning (HVAC). The cost of shipyard labour, both direct and indirect, is next. Finally there are shipyard overheads which are shipyard costs not directly attributable to the contract. Some of the indirect labour may be included as an overhead, depending on the accounting policy.

Materials costs can be estimated providing there is a specification of requirements. However as with other design information the specification will develop as the project design develops, so full and detailed information is only available at the time of purchasing. To take steel materials for a project as an example, an initial estimate of steel may be based on a previous ship. The steel quantity for the new proposed ship is estimated from the previous ship, usually by a formula using the principal dimensions for the two ships. So the estimate for the new ship steel weight would be the basis ship steel, multiplied by a function based on the ratio of the new dimensions to the basis dimensions.

Machinery cost can be estimated using a formula relating the kilowatts for the new ship and previous ships. Large items such as generators can be estimated by comparing the electrical capacity of the two ships. All early estimating methods are useful guidance, but must be used carefully.

Labour cost estimating is the main variable cost under the direct control of the shipbuilder. However because there are usually many sub-contractors and items supplied from other parties, labour generally only represents approximately 20% of the total cost of a ship. Nevertheless it is an important component of that cost.

Different approaches are used to estimate labour costs, usually single parameter estimates or time study driven methods. Single parameter estimates are generally based on past shipyard performance levels. If the past man-hours for a unit of production are known and the unit quantity for a new ship can be estimated, then the man-hours for that ship can be determined. The unit of production is usually a global measure, based on an attribute of the ship which is readily calculated at an early design stage. Typical estimating parameters to estimate the man-hours required are described here.

Man-hours per compensated gross tonne (cgt), or cgt per man year can be calculated. The compensation factor allows different ship types to be compared. Gross tonnes are available for any cargo ship. The compensation factors developed by the OECD and agreed internationally are designed to allow equivalence when comparing the man-hours required for construction of different ship types and sizes.

Man-hours per light tonne is a useful measure for non-cargo ships as it considers the total weight of the ship. It is most used for military ships.

Man-hours per tonne of steel is a factor which is most valuable for large cargo ships, particularly oil tankers and bulk carriers where the steel hull is the dominant element.

Examples of other parameters are metres of pipe, square metres of coating and there are many others, generally inverting the measure to give the quantity per man-hour. In principle, any measureable quantity can be used so the method can also apply later in a project to create detailed estimates at a local, work-station level. The availability and level of detail varies as the design develops, but dimensional parameters can generally be estimated reasonably accurately at an early design stage.

Sub-contractor costs will be based on past projects and the anticipated extent of sub-contracting for the future period. In some cases where the shipyard and major sub-contractors have long experience of collaboration, an estimate can be generated in cooperation with the sub-contractor.

The shipyard overheads are estimated at an early stage by adding a percentage to labour costs, using past information and expectations of the labour hours to be used in the period being considered.

At the build strategy stage of the contract there is preliminary design information available. At this stage a firm price is needed so that a contract can be considered. The estimate produced should be within 5% of what will be the final actual cost. As mentioned earlier, the level of design development on contract signing can vary. For the purposes of this discussion, it will be assumed that the design is sufficiently advanced for a specification which will include the major equipment and machinery items. The arrangements of the ship will be sufficiently advanced to allow the lengths of pipe and other systems to be determined with reasonable accuracy. Also the areas and volumes of spaces in the ship will be available so that estimates of quantities of materials can be estimated.

At the tactical planning stage, overall production information is available and this allows more detailed estimates to be made. Accurate resource requirements must be decided so that the project can be planned and executed. An important point here is that the total budget set by the contract price is usually fixed, so all these estimates must give a total within that budget. There may be agreed cost variability allowed in the contract where the project retains some uncertainty. At this stage the estimates should be within plus or minus five percent of an accurate figure.

The estimate determines the cost of building a ship, in a specific shipyard and at a specific time. It is subject to the variations described earlier and to internal factors in the shipbuilding company.

The price for the ship is set by the shipbuilding market which is affected by supply and demand in the shipping market. This then determines the demand for ships, also taking into account the supply of shipbuilding capacity. The capacity is influenced by possible incentives from governments, since a lot of shipbuilding capacity worldwide exists because of government intervention, both to create a capacity and to maintain it. Incentives can be offered to shipbuilders and to owners, and may not be transparent because subsidies are against trade rules in many cases.

The current and future expected freight rates also play a part, as shipowners try to anticipate market changes. Any profit from a shipbuilding contract is the small difference between cost and price and a typical return for a ship contract may be five or tem percent of the contract price.

Some further influences on cost and price include whether the shipyard has experience of a ship type. Lack of experience may result in an underestimate of the production cost. There can be a danger of the commercial imperative to secure a contract overriding the need for a realistic estimate of cost. A poor market will result in lower prices irrespective of costs, although there may be some lowering of commonly used material costs. The credit rating of the customer may influence the price which a shipyard would seek. A full order book may allow a lower overhead rate and consequently a reduced price offer.

As the design develops an accurate quantity can be determined and at the detailed design stage there is a precise definition of each steel part to be used. It must be remembered that the cost of the steel is the cost of the invoiced steel, not the steel which is actually used for the ship. There may be up to 10% wastage from part cutting and this must be added to the net steel identified from the accurate computer generated design information.

Other materials include pipes, cables and coatings. The situation is similar to steel in that detail develops over time. The preliminary estimate based on past experience and systematic variations in quantities. For machinery there will be past information probably of limited value or an estimate from potential suppliers. The same applies to other equipment.

For sub-contractors it is possible to seek quotations from potential contractors but they have similar problems as faced by the shipyard, with limited early information and an incomplete specification.

It is worth mentioning again that the actual cost of a ship may vary considerably from the initial estimate. The actual cost is likely to rise considerably for a prototype project. However, if further ships are expected to be constructed after the prototype, in a series of sister ships or at least similar ships, then the later ships can be expected to be built at lower cost. This is because there will be a learning curve effect.

Learning curves provide a means of capturing the benefits of experience. This is a concept originally developed during the 1940s in the USA. Many aircraft were produced in new factories using labour with no previous factory experience. Some means of estimating productivity was sought and it was observed that there was generally a consistent improvement. Each time the number of a product made doubled, the cost reduces by a fixed percentage. Although the theory was developed for large numbers of aircraft, the principle does apply to ships.

The estimating process, and especially the information used and generated, is a closely kept secret for a shipyard. However all shipyards use the same basic estimating models. These generally use a spreadsheet, although there is some proprietary software available. The estimate is carried out on the basis of the various material categories, because materials are the largest proportion of the total cost.

A database is often generated to provide a library of past contract cost information and current information for example on equipment costs. There are recognised categories, especially for military ships, described as ship work breakdown structures which are used alongside the product work breakdown structure. The use of materials, based on the ship systems, is logical since the bought in materials are large proportion of the total ship cost. Usually each type of material or equipment has a three digit code. For example, machinery codes may be numbers in a 400 series. A second digit identifies a particular system, for example 410 for deck machinery, 420 for thrusters and 440 for the ballast system. The third digit identifies particular items or materials within the systems, so perhaps 441 for pumps.

Typical systems and cost categories are listed here:

Steel as the main material used for the ship structure. It may be replaced by aluminium for some specialised, smaller ships.

Other hull materials, which includes paint coatings and deck coverings.

Cargo systems, which vary from a minimum on simple bulk carriers to deck cranes, container fittings, lashing systems for trailers on ro-ro ships and cargo piping systems on tankers.
Hull systems, always ballast piping, pumps, valves and other internal fittings. Deck fittings and equipment are also included.
Machinery for propulsion, including propellers and shafting.
Other machinery systems, which may be limited or be very extensive for a specialised ship.
Electrical systems with generators, switchboards, lighting and cabling.

How the various systems are designated will vary, and most shipyards have their own preferences. It is not that important provided the categorisation which is chosen is always consistent to make the data generated over time as useful as possible.

Labour is usually treated separately, with the labour costs allocated to categories which correspond to the materials as above. Alternatively the labour categories may be based on the expenditure in different work stations. The allocation will depend on the available past data and preferred practice.

Services are the final category, to take account of requirements for the contract to be bought from suppliers. Examples are design studies, model testing, specialist noise and vibration engineering or measurements using laser or photogrammetric systems.

Within the estimating model, cost may be for a quantity of materials, at a known rate. So the steel cost per tonne will be multiplied by the estimated quantity. Different costs and quantities will be applied for steel plates, profiles and any special steels which are required. Other materials will generally use the same principle. For equipment the estimate may be using a supplier quotation for a specific equipment item.

In some cases a quotation may be sought from a supplier quotation for an equipment item, though this can be problematic because the supplier price will generally depend on the state of the equipment market at the time of placing a firm order. Good supplier relations may be of help.

Variable overheads are those mainly labour costs which can not easily or accurately be directly attributed to a specific contract. The definition of these costs does vary from company to company, but it is usual to include marketing, commercial activities, stores management and services such as transport and maintenance. Others are sometimes included, such as technical staff and supervisors, although many shipyards attribute these costs to a contract. The costs are variable as they will increase or decrease with the shipyard level of activity.

The motivation is often to try to keep overheads to a minimum, as they are seen as not contributing directly to the operation of the shipyard. They may be seen as a cost rather than a benefit. So by recording the time of some of the staff mentioned above as part of a contract, the overheads are believed to be somehow reduced.

Realistically, if the work done is necessary for the shipyard operation, then precisely where the costs are allocated is not so important, as they will not go away if they are counted as direct rather than indirect. If they are not essential, then an analysis should reveal this easily and the work can be left undone. It is not uncommon for shipyards in financial difficulties to stop some work, such as maintenance and training to save indirect costs. As a short term solution this may have some effect but in the longer term it is not a good idea.

There are also fixed overheads, which are represented mainly by the costs of the shipyard facilities. This accounts for the shipyard paying for the docks, workshops and other hardware. These are all the costs that must be paid whether ships are built or not and also include rentals and local taxes in many cases. The fixed overheads are a varying proportion of the costs, depending largely on the age of the shipyard facilities. Some of these such as docks have a very long life, and especially where a shipyard has changed ownership, may have been completely paid for.

A brief summary is appropriate here. Firstly, estimating is not a precise science. It is a best estimation of the costs of building a ship, with the information available at the time of the estimate. Second, the design and other information required usually lags behind the need for it in the estimating process. Parametric methods are useful in early stages of a design, although not fully reliable so several methods should be tried. Importantly, the estimated cost of building a new ship is not necessarily related to the price that the market is prepared to pay for it.

Chapter 12
Steel Part Preparation

The fewer parts, the better. Exactly. No truer words were ever spoken in the context of engineering
—Christian Cantrell

The construction process begins with creating steel parts. Steel is ordered from suppliers according to the requirements of the ship design. Unless the shipyard is closely associated with a steel producer or supplier then their programme of steel production is unlikely to coincide with the requirements of ship construction. Especially in the case of more specialised steels, it may be necessary to order steel and have it delivered very early. It is usual to have a stock of steel available even for more standard products, to ensure continuity of production if there is any interruption to the supply. Steel is stored until required, then cleaned, straightened and usually coated with a primer. The parts are then cut from the steel plates and profiles, and bent to shape where necessary.

The first stage in the process is the stockyard for steel. The need for this is dictated by the ability to match supply of steel to the facilities with the demand. If the steel mill is close, and can provide steel with a very short lead time, then storage is not a problem. However as a precaution, most shipyards have a steel stock for one to three months and a longer period for specialist steels to ensure availability of the material when required.

Plates are generally stored horizontally and handled individually by magnet beam cranes. This makes the handling faster and safer, and makes sorting the plates into specific locations according to production needs easier. Profiles are usually handled in batches using slings for speed, although magnet beams are also available for individual profile movements.

For smaller ships, the use of standard sizes of steel plates is more common and this reduces the need for storage. If standard plates are used then it is only necessary to take the top plate off a pile and no sorting is needed.

For larger shipyards with a high throughput of steel, the movement of steel from the storage area into the preparation workshops is usually by means of a roller conveyor. This allows steel plates and profiles to be loaded onto the conveyor when the crane is available, and then the several items on the conveyor create buffer storage of material

G. Bruce, *Shipbuilding Management*,
https://doi.org/10.1007/978-981-15-8975-1_12

ready for the preparation processes. Up to two hours worth of material may be stored before the next set of plates or profiles are loaded leaving the crane available to unload incoming transport and sort material ready for loading on the conveyor. The objective is to allow efficient use of the crane and at the same time to ensure that the preparation processes are not held up for lack of material.

After the storage and prior to preparation, the steel is usually shotblasted and primer coated. In some cases for smaller throughputs of steel and for smaller ships, the shipyard may choose to purchase the steel ready blasted and primed. This avoids the need for expensive treatment equipment as outlined below. The decision is economic, so is dependent on the cost of bought in as opposed to shipyard treatment. In a few cases groups of shipyards operate a single treatment facility, serving a number of companies. The logic is that they all require the service, and with a sufficient throughput of steel the cost is lower than if bought in ready treated. The treatment of the steel is not part of any competition between them.

The treatment process includes a number of work stations, all automated. These are pre-heating, usually water wash or gas, plate levelling, shot-blasting, primer coating application and drying. The processes are linked by a roller conveyor, and the operation is largely automated.

The steel first passes through a cabinet with water wash or gas pre-heating equipment. This raises the temperature of the steel which will speed the drying of the coating to be applied later. It will also partly clean the steel, removing or loosening mill scale and other debris.

The next cabinet contains shot blasting equipment. As the steel plate passes through, there are impellers which throw steel shot or other abrasive onto the top and bottom surfaces. These impellers work transversely across the plates, ensuring complete coverage. The speed of transit of the plates can be varied to give a clean surface which is clean for processing and suitable to be primer coated.

A paint cabinet is next, again with in this case spray heads passing across the plates as they transit the cabinet. From the coating, the plate passes through a tunnel with fans to blow air across the coated surfaces. In combination with the higher temperature following pre-heating, the primer coat is dry enough to allow the plates to be handled after they emerge. Further conveyors in the line then provide buffer storage and often distribute the plates to the correct work station for further processing.

A separate treatment line is used for profiles, being designed to manage their various shapes. This has the same processes except for levelling.

Steel cutting is the next process. Plates are generally cut using numerical control (NC) machines. In the past some older shipyards used optically controlled machines but the availability of basic NC machines and associated computer aided design packages allows these to be used almost universally.

Various processes are available for cutting steel plates and profiles. This is a major feature of large scale shipbuilding, and important for all shipbuilding because the starting point for efficient production is the availability of accurately cut parts.

Oxy-fuel torch cutting, or flame cutting, is the oldest cutting process that can be used to cut shipbuilding steel. It is used in smaller shipyards and where welding preparation is required. The process uses a heating gas, usually propane, and oxygen to heat the steel, and then a second oxygen feed to blow away the molten steel leaving a cut. The path traced is managed by the numerical control system. It is a simple process, and the equipment and consumables are relatively inexpensive.

An oxy-fuel torch can cut any thickness of steel used in the industry. Careful adjustment and regular maintenance of the equipment will give a smooth, square cut surface. It is not uncommon in shipyards for there to be a lot of visible slag on the cut plate edge. By using two or three cutting torches together, the process can cut weld preparations for thicker plates. A single cutting machine can have a number of cutting torches on it, allowing several parts to be cut simultaneously including port and starboard shell plates with the correct control equipment.

Oxy-fuel cutting is suitable for all shipbuilding steel, though steel plates which are thinner than 8 mm in particular are likely to suffer from distortion due to the heat generated from the cutting process. The process is relatively slow, up to 500 mm per minute being the fastest cutting usually obtained.

Plasma arc cutting is often the favoured process in shipyards, because it cuts at much higher speeds than oxy-fuel cutting. The edge quality is poorer, and the edge may not be square, although it is generally acceptable for shipbuilding steel in the range 6–25 mm, which covers most requirements.

Plasma equipment is more expensive than oxy-fuel, but the higher cutting speeds available will make this worthwhile for large scale production. Plasma is usually limited to one cutting head, although two or more can be fitted to a cutting machine. The high cost makes plasma viable for large production volumes. It is essential to carry out a full analysis of the cutting requirements in a shipyard, taking into account plate marking as well as cutting. The materials handling of steel, especially removal of cut parts, is also important and if done inefficiently can reduce the cutting capacity. Work study and sometimes simulation can be used to analyse the whole cutting operation.

For plates the usual solution is a cutting bed twice the length of the longest plate to be cut. This allows one plate to be cut, while the parts from a previous plate are removed and a new plate is aligned on the cutting bed. This keeps the machines in nearly continuous cutting. In a few cases conveyors are used to move materials into the cutting area and parts to storage after being cut, but this is expensive.

Laser cutting is available, but is rare to date and only generally used for thinner steel plates, for example superstructures and small military ships. It is suitable for cutting shipbuilding steel up to around 25 mm. but is limited to a single cutting head on a machine. The edge quality is usually good. The major benefit of laser cutting is the very narrow cut width, and the speed of cutting which minimises the heat affected zone of the plate. It is also a fast process which also reduces the cutting time of the plate. If the steel plate is heated excessively, then stresses in the plate can be released as the plate cools, leading to distortion which can have a serious effect on part and subsequent assembly accuracy.

Whatever the process, the machines are usually computer controlled and are also able to mark the plates, using a powder injected into a flame without the cutting oxygen, or a defocused laser beam.

Other cutting systems found include mechanical planes which can cut straight edges only, but are very accurate. Edge milling also only produces straight edges, and is very slow but might be valuable for very specialised weld preparations. Both of these are usually old machines which have remained in a shipyard after modernisation. Guillotines are effective for small, straight-sided parts such as brackets which are required in large numbers.

Profiles are cut using plasma, oxy-gas systems or mechanically. Robots can be used for high volumes of throughput. Mechanised conveyors may be used to move the profiles from storage to the equipment and then on to the storage for the next process. Hand or portable equipment is still used in small shipyards. As with plates, a very careful analysis of the requirements for parts and the process from storage to removal of cut parts is needed to ensure any investment in automation is justified.

Steel plates for the curved sections of the ship hull have to be formed to give them the correct shape. The plates are formed using a number of machine types. For small volumes of production, some of the machines can be used for other purposes, such as swedging plates. Heat-line bending is an alternative, but slower and labour-intensive. Some shipyards use a mechanical forming machine for curving plates, and then use heat line bending to increase the accuracy of the final product.

The machines which are used to form steel plates include a plate rolls which is used for shell plates, and is best for single curvature. Plates such as bilge plates near the midship section are examples. Where a double curvature is needed for example towards the bow and stern of a ship, a ring frame press can be used. A portal press can also be used for most forming requirements, using a variety of forming tools for specific shapes. However the process is also slow because the plate is shaped one small section at a time. A press brake is used for flanging brackets and other minor structure which does not then need welded stiffeners. It is also used for corrugated stiffening, again to avoid welded stiffeners for example for bulk carrier bulkhead sections.

Heat line bending relies on the deformation of a steel plate when heated and then cooled. Whereas this is a problem in cutting and welding, as the resultant distortion causes problems with accuracy and fitting parts together, it is useful for forming plates. A gas torch is used to heat the steel, moving along a pre-determined line marked on the plate. Water may be used to cool the steel quickly. The effect is to cause the steel plate to bend along the length of the heated line, inducing a curve. Repeated applications are used to achieve the final shape so it is a slow process.

The shape is derived from the CAD design, and it is usual to use the information to create a template for the required curved shape. This is used to check the curvature until the correct shape is reached. Accurate mechanical bending is difficult because of "spring-back" so this is another reason to use heat line bending is used for final adjustment of formed plates.

Profile forming is primarily for transverse frames and uses a frame bender. This machine can hold two frames, port and starboard, at the same time and they are formed together. As with plates, the templates are defined by CAD. The final shape of the frame is checked using one of several methods. The most effective is called inverse-curve bending. The frames are marked with curved lines, derived from the

CAD system. When the frames are the correct shape, the marked lines become straight, which makes them easy to check and this assists with accuracy. Wooden templates, or wire sets bent to the correct shape can also be used to check the shape of the frames. These are offered up to the frames and the curve adjusted until there is a correct fit between template and frame.

Other processes can be used for flanges and small stiffeners, including a press brake, which is also used for flanging brackets. A mast rolls is an open ended rolls which can form plates into a complete circle.

For smaller shipyards, there are sub-contract companies able to carry out all the cutting and forming processes and deliver parts ready for use. As in the case of plate and stiffener treatment, this can avoid the need for expensive, fixed investment which for a low throughput of steel will probably be underutilised. Also as is the case with steel treatment small shipyards have sometimes set up a company which carries out the processing for all of them. The production of parts is also not seen as a competitive issue, compared to assembly and especially outfitting.

In many shipyards, part production for outfitting is less advanced than steelwork. There may be automated or semi-automated processes at early stages but the processes in general are manual. Outfit assembly is often limited in scope and manual. The reasons are largely due to the wide variety of equipment and systems to be managed.

All ships require a set of pipe systems which can be extensive. A pipe workshop is therefore a significant element of the shipyard facilities. In a larger and advanced shipyard, this will be a considerable investment. Pipes are delivered to the shipyard and then moved, usually by vehicle but sometimes using a conveyor, to a storage location. Generally the pipes are stored on racks and a crane loads and unloads these. In some shipyards, a large fork lift vehicle is used and also to deliver the pipe to the first process.

The equipment found may include automated storage and retrieval of pipes where production volume is high. Like the system described above, this has a series of racks for the pipes, each rack with a gentle slope from the input side. The pipes are loaded onto the racks depending on their diameter, wall thickness, material and other specifications. When a pipe is required in production it is collected from the rack using an automatic device with forks which picks the pipe from the rack and lowers it onto a conveyor. The conveyor moves the pipe to the correct process.

The first production process is generally pipe cutting to length. This may be automated for large production requirements, in which case the in-feed conveyor above will move the pipe into the cutting system. The leading end of the pipe is located against a moveable stop which has been set to the correct pipe length. The pipe is then cut using a band saw or hot cutting process, which can be oxy-gas, plasma or occasionally laser.

Smaller shipyards can use more manual methods, handling pipes by vehicle and crane onto a work table and using more basic measuring methods. A band saw is the usual cutting process.

Cut pipes then have flanges welded to their ends for future joining into complete pipe systems on the ship. Automatic flange welding can be used, where the pipe is located in the equipment, then the flange is positioned against the pipe and welded as the pipe and flange rotate, keeping the weld head still. The positioning of the bolt holes in the flange can also be done accurately so that they line up for pipe joining. The investment required has to be justified by the ability to produce many pipes quickly, with a lower cost than using manual methods.

Alternatively, the whole operation can be manual, with the pipe positioned on rollers so it can rotate, again to keep the weld in the same position rather than trying to move round the static pipe. To simplify the process the flange can be allowed to rotate about the pipe, being held on the pipe by a narrow flange which will butt against a similar flange on the adjacent pipe in the system. The rotating flanges are easily aligned when the pipes are joined.

Once the pipe is flanged it may need to be bent. Ideally the pipe system design will minimise the number of bends by running pipe systems as far as possible along pre-determined straight paths. Some bending is still going to be a requirement. For large diameter pipes the equipment is a major investment, in particular if the shipyard decides to use a numerically controlled system.

The pipe bender requires considerable set up time when changing from one diameter to another. Changing the radius of the bend is also time-consuming. As a result the scheduling of the pipe bending operation requires careful organisation, otherwise the machine can spend a lot of time in set up and less in productive bending activity.

Other specialised equipment is used for branches in pipes, using computer control to cut the correct holes in a pipe so that a branch pipe can be joined to it at an angle. The use of design for production can reduce the complexity of pipe systems in many cases, essentially running pipes as far as possible in predetermined straight lines. Bends and angled branches are eliminated if at all possible, which can eliminate the need for more investment.

Pipe assembly is generally a manual operation, with the pipes located in a jig on a work table. One the pipes are located and checked for accuracy they are tack welded and then finish welding is completed.

Pipes can be assembled from bought-in bends and straight pieces, which can eliminate some of the cutting and bending equipment and therefore reduce the investment cost. This is usually where only a small number of pipes are required, so the investment in equipment is not cost-effective.

Other outfitting workshops generally provide basic facilities if the work is not sub-contracted which is the usual practice for current shipyards.

Sheet metal work for ventilation and metal furnishings requires small mechanical cutting and forming equipment. Lock forming for ventilation trunking and other specialist equipment may be required. It is usual to complete the work by use of manual assembly.

A machine shop was a feature of shipyards which often made some of their own equipment. Also the workshop was used for making specialised parts for fitting, bolting down and aligning equipment. It is more common now to buy in equipment, often ready mounted on supports. Resin chocking has eliminated most of the alignment requirements. Where a workshop is provided, it is generally equipped with general purpose machine tools. Specialist equipment, for example a shaft lathe for propeller shafts may be available. These are often old equipment which like some of the steel plate forming machinery has a very long life if well maintained.

Accommodation on ships was traditionally high quality woodwork, but is now almost always bought in from specialist suppliers, and in non-flammable materials. Bathroom units are generally bought as complete modules to be installed as a single item. For most commercial ships, the number of crew is small compared to the past, so the accommodation needs are reduced and investment in any facilities, except for the very largest shipyards, is uneconomic. Passenger ships are an exception where large scale accommodation is required. Most builders then use specialised sub-contractors who can produce in a factory environment and at lower cost.

In analysing the part production processes, and the associated handling, work study charts can be useful. These are discussed in the next chapter, but one specific technique is very useful here. This is the operation chart, also known as a "right-hand/left hand" chart. It is used for very detailed analysis of a single operator, or for a single machine. It is only used for manipulative operations with short cycle times and high volumes of throughput. These may be locating, picking and positioning parts, or moving a tool to complete a process task.

The technique is usually associated with detailed analysis of an individual worker, often on a production flow line. It is also a basis for the development of robotic operations where the activity must be defined in detail.

Automation is of considerable interest to shipbuilders, especially those with advanced management and facilities, building relatively large ships. A question to ask is what processes in shipbuilding are suitable for automation? It is included in parts manufacture as this is the area where automation is most easily put into operation. Later assembly stages of the ship construction project result in large structures, often with difficulty of access and so automation can be much more difficult to implement. A further question is what justification is there for automation in marine production. It is expensive to bring in automation and it results in higher fixed costs compared to the relative flexibility of using more labour intensive methods.

As a start then, there is no universal definition of automation. Often what are described as robots are more simple mechanised aids to production so it is useful to consider various stages of development from purely manual work to full automation. One definition of full automation is "The carrying out of a process without human intervention."

This immediately raises several questions of which the first is when does the process start and finish? It can be considered from the initial movement of input materials to a process, setting up the process, working, inspection, removal and preparation for onward transfer. Not just a robot for example to weld steel parts, but also the whole necessary infrastructure to make this efficient.

Then the next issue is what human intervention is permissible if a process is to be considered to be automatic. It is useful here to consider the components of automation. Any industrial operation can be considered to have four components. These are processes, materials handling, control and mechanisation. To take a very simple example, a person can use a hammer, the process, moving it to the work piece, controlling by eye and hand and using human muscle rather than anything mechanical.

By processes is meant the tools and equipment which are used to perform tasks. These can be anything from the hammer to an advanced welding process, simply the device that carries out the work operations.

Materials handling is the transfer of materials between processes, the storage between processes and the manipulation of materials to allow the process to be performed efficiently. All these aspects of handling can occupy a great deal of time and effort in a shipyard, but are essential to efficient production.

Control of the process governs the regulation of both quantity and quality of outputs. At the simplest level, hitting the correct part with the hammer and using the least number of strokes.

Mechanisation of processes can be used to reduce human effort and introduce power technology, which is another source of energy for the activities. So a hammer can be mechanically operated to give more power.

Each of these four components can be considered to be on a scale from purely human to fully automatic. Examples can be found including, manual processes, with materials manhandled to where they are required, controlled by humans. These are basically hand tools. Then there are power-assisted processes, for example an electric drill, which can be mounted on a bench.

Hand-controlled machines can then be used to improve the materials handling, for example a roller conveyor to take work to and remove products from the work process. This is under human control but can be extended to have automatic delivery of items to the operation.

Automatic machines, which can be hand activated, such as an N.C. plate cutting machine can be used. The processing is automatic, but cranes generally move material in and out, under human control.

Fully automatic processes such as a robotic cutting machine can also be used. The robot will be fixed in position, so roller conveyors are used to input and output materials. These have product detection, for length measuring and accurate location.

The applicability of automation generally depends on the number of products required in a given time, and on the variety of products. Standard products can be made using "hard" automation, dedicated to a single or very few items and not changing. Robots or other automated devices such as NC cutting equipment are applicable where various different products are needed, though the products will all be of reasonably similar size and configuration.

Robots are used in many industries. The term robot is often misapplied to simply automated devices, or tele-operated equipment. There is no universal definition of a robot available. One robot industry association definition is a programmable, multi-functional manipulator designed to move materials, parts, tools or specialised devices through variable, programmed motions for the performance of a variety of tasks. The key element of the above is that the device is programmable and so adaptable to alternative tasks.

Robots are adopted for several reasons, first to overcome shortages of labour, either an absolute shortage of people, shortages of skills, or a poor image of industry.

Second, robots are used to improve productivity. They can also carry out tasks in poor working conditions. Robots offer flexible automation, with just-in-time production and smaller batch sizes. However for ship production, inhibiting factors include the generally large work pieces, fixed position work pieces which are difficult to manipulate and generally difficult working conditions.

A cost–benefit analysis is essential for robots, or any other item of equipment. It is also important to consider that a robot requires a lot of support, for example handling work in and out of a process, manipulators to make the work accessible and usually a good environment. In many cases it is relatively low cost to provide the support but expensive to provide the robot. So for shipbuilding, it may be better to address the support features first because a significant improvement in conditions and productivity can be gained for a small investment.

Chapter 13
Assembly

Most of the time spent trying to make new technology work is great fun for nerds, but not worth it for the end users
—Douglas Adams

The main element in more advanced shipbuilding is assembly. For a well developed shipyard, using the most efficient processes, the assembly activities consume the majority of steel work man-hours. Parts are made, but can almost entirely be purchased from suppliers as an alternative, but assembly largely defines the industry. So shipbuilding is primarily an assembly industry, using the parts created in previous processes.

Considering first steelwork assembly, this is usually carried out in several stages and many local definitions exist for these. The simplest general definitions are:

Assembly, which is a generic term used to include all stages where interim products, from parts to small assemblies, are joined together.

Minor assembly is a sub set of assembly and describes the joining of two or more individual piece parts to form a small interim product. Typical minor assemblies are brackets with webs and stiffeners, stiffened floors for double bottom units and side frames.

Sub-assembly describes the joining of parts and minor assemblies to form a larger interim product, for example a double bottom girder.

Unit assembly describes the joining of interim products to create a unit which could be taken to the final construction site for installation on the ship. A unit is the smallest steelwork assembly that it is convenient to install on the ship. This is because the objective of unit assembly is to reduce the work content of the installation and construction of the ship in the construction site, whether a dock, slipway or other berth.

Block assembly describes the joining of units and sub-assemblies to create larger interim products before they are taken to the construction site. The larger blocks are a means of further reducing the construction man-hours.

G. Bruce, *Shipbuilding Management*,
https://doi.org/10.1007/978-981-15-8975-1_13

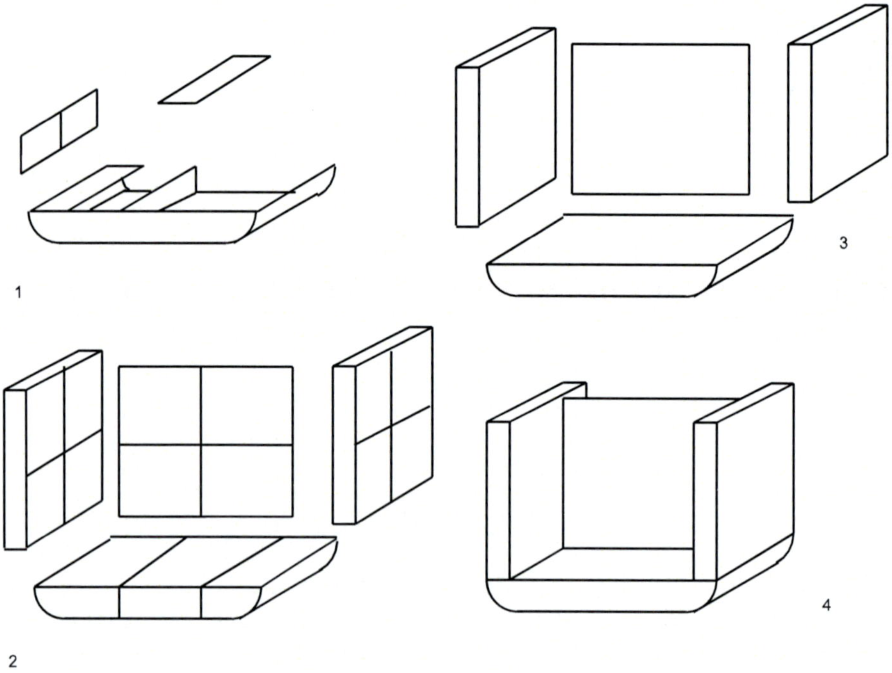

Fig. 13.1 Ship construction development

In a small number of shipyards the blocks may be combined prior to incorporation in the ship into complete ship cross sections or rings. These may be of sufficient size that they are not lifted into position but moved on ground transport.

Figure 13.1 shows four stages of assembly in the context of the ship. Figure 13.1(1) shows sub-assemblies, including stiffened shell plates, bulkhead plates, double bottom floors and girders. Figure 13.1(2) shows examples of units, including bottom units, side units and bulkheads. In Fig. 13.1(3) the units have been assembled together to create blocks, again with side blocks, bottom blocks and bulkheads. In Fig. 13.1(4) the blocks are further assembled to create a cross section or ring, basically a complete cargo hold.

Specialised equipment is often used in the assembly stage. Flow lines are created for some of the interim products, including some flat sub-assemblies and unit assemblies. The first such lines were produced for production of flat panels. These were a result of rapidly increasing ship size, particularly oil tankers, in the early 1960s. In order to construct the ships in a sensible timescale, large unit were required, and because of the hull shape and construction, the units were largely based on flat, stiffened plate panels, typically deck, side, bottom and bulkhead units.

The flow line for flat panels comprises a series of discrete work stations, each with a specialised function, frequently a flow line with work stations linked by conveyors. The conveyors move the panels from station to station without the need for large cranes, and consequent heavy duty building structures. The panel line has specialised equipment for each stage of assembly. These vary a little depending on the panels to be produced, but a typical set of work stations is as follows.

First, steel plate alignment and fairing using magnets. The plate panel is assembled with the plate seams over a double line of electro-magnets, which are energised once the plates are correctly positioned to hold them in place. The plates are then tack welded to preserve their accuracy.

Second, plate welding, which can be using single side welding equipment, or using a panel turnover crane using double sided welding. The latter is simpler in terms of the welding requirements. Single side welding requires less investment in building and eliminates the need for a crane capable of lifting and turning over the panel. On the other hand single side welds may require some weld repair work. For thicker steel plates several welding heads are provided so the plates can be joined in a single welding pass.

Marking of stiffener positions is sometimes the next process, although this may be done at the plate cutting stage if the shipyard has invested in good quality management procedures. In the same or possibly a subsequent work station, the stiffeners are located in their correct position on the panel, held in position using hydraulic rams mounted on a steel frame over the panel, faired and then tack welded into position.

The fourth station is for stiffener welding. This is a high speed, usually multi-wire, submerged arc process which completes the fillet welds for both sides of the stiffeners in as short a time as possible. More than one welding system may be used, because this is usually the potential bottleneck work station which can set the frequency of panel production for the whole line. Alternatively a second welding station is provided on the line, to ensure the time taken to complete work at this work station is as near as possible the same as for the other.

Next, the secondary stiffeners are located in position and tacked. These may be substantial sub-assemblies such as girders and webs, for double bottom structures. Usually there is no specialist equipment to position these stiffeners, as their geometries and dimensions can vary substantially. There may be two work stations for these processes if the structure is complex.

The welding of the secondary stiffeners at the next station is as automated as is possible but this does depend on the type of structure to be produced. Generally automated welding systems are preferred and in some cases robots are in use. More than one welding system may be used to maintain the necessary production rate.

Once a panel is complete, it is inspected to ensure there is no outstanding rework left for the next stage of production. Where the space on the completed ship will be outfitted, for example cargo piping for an oil tanker, some of the work can be done at this time, with the panel is in a convenient orientation and in a workshop. Pipe supports can be welded in place and the pipes may also be installed. The panel can also be coated prior to the next stage of work.

For double bottom and double hull structures, a further work station may be included where one stiffened panel can be turned and positioned on a second to create the double panel. Welding is necessarily then in a confined space and there may be several static work stations for the purpose. Installing pipes and coating the internal spaces is even more important in these cases.

Clearly even the largest ships with high block coefficients have curved parts of their hulls. Another specialised set of workstations can be found for the assembly of curved panels. There are two options.

The first provides an accurately shaped jig for panel assembly. Built on a level base with a grid of steel beams on the workshop floor, the workstation has a set of telescopic pins of adjustable height. A base tube is welded to the floor grid, then a second tube slides within the first so the height can be adjusted. Often a vernier scale is used for the positions of holes in the inner tube, while the outer has holes at fixed spacings up the tube. This allows the height to be adjusted within 50 mm of the required height. A screw adjustment on top of the inner tube allows a final, accurate height to be set.

Magnetic or other fairing equipment is used to align the plates (which will form part of the hull on the jig). The internal stiffeners are then added and welded. Single-sided welding is used to avoid working under the jig as much as possible. This approach avoids the need to turn the finished panel. Using the pin jig, errors can be created in the jig heights, from plate forming and in the fairing of the stiffeners. Depending on the levels of accuracy achieved in a shipyard, which is a function of the quality management, the best choice of assembly process can be decided.

The pin jig is likely to offer good productivity if the accuracy is available, but an alternative is to assemble the internal stiffening as a matrix on the floor grid. The plates are then lowered onto the stiffening and faired, then welded. There is a need for overhead tack welding and then the tacked panel can be turned over for final welding of the stiffeners to the panel. Using the internal stiffening as the jig makes the fairing easier and as long as the accuracy of the parts is certain, gives a very accurate result.

The next main stage of assembly is unit assembly. Several sub-assemblies, for example two or more panels, which may be flat or curved, are joined together. This is normally done in a large workshop equipped with high capacity cranes. The work can also be carried out near a dockside, from where the construction cranes can lift the finished unit to the ship if no further assembly is required. For a small shipyard the construction cranes can also be used for the unit assembly work.

Unit assembly is almost always carried out in a fixed position, with a first sub-assembly located and levelled on the floor then others added and tacked in position to complete the structure. Then the assemblies are welded together. The unit may require to be turned through ninety or one hundred and eighty degrees during the assembly process, using the workshop cranes. This ensures it is in the best orientation for the fairing and welding operations. The welds are generally completed before the unit is turned over. While the unit is in its initial position, it is often useful to carry out further outfitting, as a minimum the installation of pipe supports, cable trays and other foundations for fittings or equipment. It is particularly useful to finish work which will be on a deckhead in the ship, or high on a bulkhead, to avoid some scaffolding at a block stage or on the ship.

Further block stages may be found for very large ships, and for smaller ships in some cases. Several units are joined to form a block, which might be a complete double bottom structure for a cargo hold, a complete side unit including part of a cargo bulkhead or a large deck unit. This work is more likely to be carried out beside the shipbuilding dock, to make use of the cranes, but for large scale production a further specialised workshop may be provided.

Advanced shipyards combine steel and outfitting at the earliest possible stage of production. It is more usual to install equipment and other outfitting at the block stage. Typical activities carried out on completed hull blocks are shotblasting, paint coating and outfit installation. Shotblasting is performed to remove any corrosion products from the steel surface. Some steel in shipyards is coated with a primer before cutting and assembly work, to protect against corrosion. This coat is often damaged by the hot processes used and by abrasion so is removed at the block stage if it cannot be repaired.

The work of shotblasting is often sub-contracted as being of relatively low value to the shipyard. In such cases the contractor will operate the facilities provided by the shipyard. A serious part of the work is cleaning the shotblasted block after completion of the process. Vacuum systems are used as well as shovels and for a large complex block the issue is serious. Under floor conveyors are sometimes used to remove spent shot and debris which is cleaned off the structures. This shot is then filtered to remove the old paint, corrosion products and other waste, and the useful shot recycled for further use.

The final coatings are then applied in paint cells. These are specialised facilities with temperature and humidity control and good access equipment. The buildings have to be large enough to accept the largest blocks assembled in the shipyard, although unlike assembly buildings they do not require any significant cranes. A range of methods is available, even including brush or roller for areas with poor access. Usually airless spray painting is in use.

The vehicle design has to take into account the need both to facilitate coating and to avoid corrosion. Design can offer simpler structures, which can enhance assembly as well as coating quality. A good design can help to minimise the work done at late stages when coatings may be present.

Important considerations in cleaning and coating include the need for good access to allow for thorough cleaning, for surface preparation and subsequent removal of spent abrasives. Areas where abrasive can be trapped should be avoided as far as possible. General work access is important to avoid the need for staging, for use of hand tools and more especially for automated systems. The integrity of the final coating is important and it is necessary to smooth any rough edges on the steelwork which might be a starting point for coating breakdown. This is usually by grinding the edges and is done throughout the assembly process, when it is most convenient.

Shipyards may also have large outfitting halls with good access equipment and cranes. These provide an undercover work space for the final outfitting of large structural blocks for the ships. Although the access requirements are more onerous at the block stage than earlier in the assembly processes, competing outfitting here will still provide a time saving during the dock construction.

It is essential to design the assembly processes. This may seem obvious, but past observation of assembly workshops in many shipyards indicates that it is often done poorly or neglected. Techniques for analysis of the processes and their potential improvement are discussed later, but it is worth considering some examples first.

In many shipyards much effort is spent and man-hours consumed, in picking up items and placing them elsewhere. This can be transporting parts and assemblies from one process or workshop to another, turning or other re-orientation. There are also lifting, manipulating and locating items ready for assembly. In some cases the movement is merely sorting, for example taking the correct piece from a pile. It can also be as basic as placing an item on the floor after a process is completed to wait for transport to the next stage of work.

Much of the effort used for these movements is non-productive. As an example consider profiles which have been cut and prepared for assembly onto a flat panel. From the preparation processes they may be lifted by crane and placed on the floor of the workshop, in a storage area to wait for transport. When the transport arrives they will be lifted onto the vehicle, and this may require a worker to assist the crane operator to position the profiles so they can be picked up.

They are then taken to the assembly input storage area, and unloaded by crane onto the floor. When required they are lifted onto the plate panel, which may require sorting them into order and a worker or two to assist the crane operator. The crane will then locate them on the panel where the workers will position them accurately then tack them into place. Each profile is handled individually.

Alternatively, once the profiles are ready in preparation, they can be loaded by crane onto a pallet, which has slots on its upper surface. The profiles are held upright in the slots and they can be placed in the order in which they will be fitted to the panel. The pallet is then transported to the assembly workshop and placed in storage. When the correct panel is ready, the pallet is placed adjacent to it and the crane can lift the profiles without assistance and locate them on the panel. A support placed on the panel can be used to keep the profile upright, ready for welding to avoid tacking.

Overall fewer workers are required to intervene to assist the crane operator and each profile is only handled once between preparation and assembly. The location of the profiles during assembly is also easier and faster so there will be cost savings from relatively simple adjustments to the assembly operations.

Movement of materials in shipyards is a major element and can be a very large consumer of man-hours. A well-designed layout for the shipyard and careful analysis of movement requirements can reduce the costs. However the scale of the shipyard, size of many items and the need to move all the materials, incorporated in assemblies or not, to the final ship construction site, mean the transport will always be an important item.

The workshops are generally equipped with overhead electric travelling cranes, with varying capacities. Up to 200 tonnes is not uncommon for large ships, and two cranes can work together for a heavier lift. Smaller cranes, up to ten tonnes, carry parts and smaller assemblies. The smaller cranes are often on rails at a lower level so they can move underneath the main heavy cranes.

Cranes can position parts and assemblies onto larger structures, so are versatile. They can also lift items onto taller assemblies and over obstacles. Most cranes have a smaller auxiliary hook, usually around one tenth to one fifth of the main hook capacity, which is faster than the main hook for use with smaller loads. The cranes are usually able to travel the length of the workshop on rails mounted on beams within the workshop structure. Some are portal cranes, running on ground rails, which is an alternative to two levels of overhead rails.

Final assemblies prior to removal to the ship construction site can be moved by using the cranes on the building slipway or dock, but there is usually a need to store a number of these assemblies prior to their erection onto the ship, and it is practical then to use a large, self elevating transporter.

Work study is a tool to assist in the process of achieving efficiency. It is applicable in a number of circumstances, including the design of production processes, for periodic reviews of the efficiency of a process or processes as a starting point for improvement and when required for trouble-shooting. It is a method for the analysis of operations and is performed by initially breaking down work into a sequence of linked elements, then reviewing each element to seek efficiency. Finally work study combines the elements into an efficient process design.

Work study charts are one of the key techniques which can be used. Work measurement is the other key technique. Work study charts are a specialised form of network which is used to create models of actual or proposed operations. There are a number of commonly used types of chart. These are process charts, flow diagrams, assembly charts, multiple-activity charts and operation charts. Work is split into operations, inspections, movements, delays and storage, with five symbols to represent these.

In using a process chart the chronological steps of the process or activity are listed, using the symbols. The steps can be in terms of different participants in the activity. The first of these are the worker(s) who can be engaged in an operation to make something. However the workers in most production are never solely engaged in operations, unless on a very well organised production flow line making small parts. They may also be engaged in transportation, of raw materials, interim products of finished items. In many cases the workers will also have to inspect items during the production activity. They may be delayed, taking an unscheduled break, because of a shortage of material or parts or equipment. Finally they may be in storage, the term being usually applied to materials but in the case of the workers representing planned interruptions in production such as scheduled breaks.

The product or interim product or raw materials may also be in different states. They can be in storage, undergoing an operation, being moved to a next operation or being inspected before further work is done. There may be delays, where the items are waiting for new parts, or for a worker to become available, or for some equipment to be provided so that work can continue. Any equipment needed by the process under consideration can also be in the different states, either in use, being moved to a new work location,, waiting for inspection, or for new material to arrive and it can also be delayed by a shortage of materials, or workers. Equipment is often also in storage.

Figure 13.2 shows a sequence of work. First the worker moves to where his material is in storage. Then the material is inspected to ensure it is the correct item, and is ready for use. There could also be a recording operation at this point which would delay the material. The worker then moves the material to his work station, and

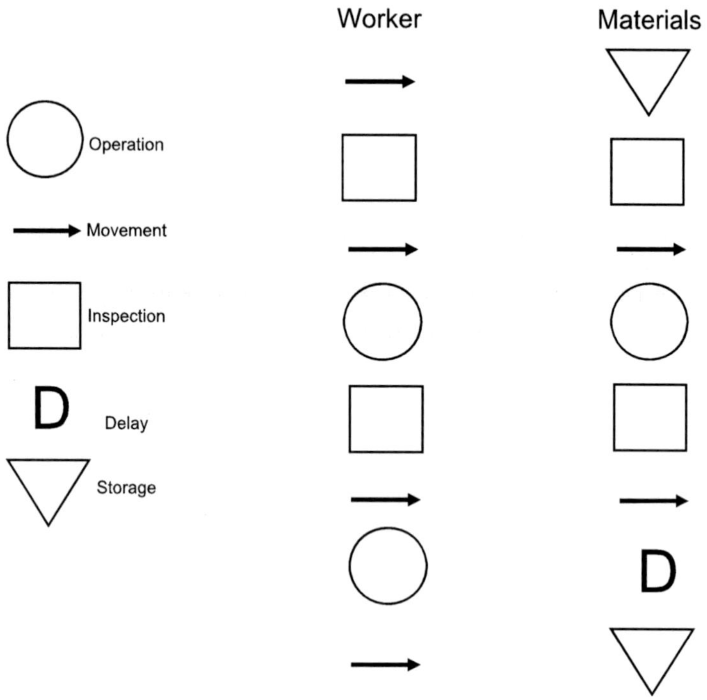

Fig. 13.2 Example work study chart

performs the necessary operations on it. The work is then inspected and found to be correct so is moved to storage for the next worker to collect. At the storage location the worker records the completion of the operation so the material is delayed. Finally the material is placed in storage and the worker moves to find the next task.

While this is very simplified it shows the basics of the chart.

Having created a chart, then the times taken for operations, movements and so on can be added. Also distances that workers or materials have to move can be included. The process chart can be analysed to seek ways to eliminate unnecessary work especially movements to combine elements, for example a movement and a delay using a conveyor for example. It may be possible to change the sequence of work to make an improvement.

There are different types of process charts, for example it is common in shipyards to use an outline process chart which shows only the operations, other elements being subsumed in the operation times. An assembly chart traces the sequence of assembly from individual piece parts through stages of sub-assembly to final product assembly.

A flow diagram places the process chart on a floor plan to chart the flow of work, which can identify bottlenecks, excessive travel distances and unnecessary movements, and provide information to relocate equipment. Once a process has been analysed or designed and a workshop layout developed, the future operations can be the subject of a simulation study. This takes the proposed operations, uses work measurement if past data is not available and then models the probable outcomes.

The workshop arrangement can then be refined to try and achieve the best possible production output.

Work measurement is usually associated with detailed analysis of an individual worker, often on a production flow line. The technique is also applicable to larger scale operations, so can also be used in the shipbuilding industry. It is used to determine a standard time for operations, although where some industries work with seconds, in shipbuilding it is more usual to find minutes or hours.

A number of assumptions are made when work measurement is used. First it is intended to consider normal workers, of average strength, intelligence, speed and other characteristics. Next the workers are assumed to be using appropriate methods for the tasks being carried out, and also to be using appropriate machinery and equipment. This immediately presupposes that some work study has been carried out and the activity being studied has been carefully designed to allow the appropriate equipment and other elements of production to be used. If this is not the case and workers are improvising because suitable equipment is not available then the work measurement is of no value. Basically the process under observation would not be in control in quality assurance terms.

Provided the workers are acting appropriately, then the next assumption is that they are working with normal skill and effort. This presupposes suitable training, and again this may immediately point to a need for some additional training before the study continues. If this requirement is met, then the workers are allowed to take normal breaks, which may be dependent on the working conditions. Longer breaks would be permitted for poor conditions, for example overhead work or confined spaces. There may also be health and safety needs associated with the working conditions in extreme cases.

Finally allowances are made for worker fatigue, for example towards the end of the working day. Further allowances are made for machine adjustments, for cleaning (for example of weld spatter) and any other specific requirements for the activity.

Work measurement is a basis for planning and estimating in particular where the company is planning to change parts of the production system in order to make improvements. It first provides an accurate picture of the current performance of work stations, and on the basis of improvements planned or after there implementation, what the future performance should be.

As such it allows planners and schedulers, supporting commercial and production functions, to determine the labour required for the shipbuilding programme, actual and planned. It also allows them to determine the other production resources needed, from significant changes to additional small-scale production aids.

It us also possible and important to make allowances for any non-productive time. Going back to the work study charts this can be because of delays in supply, also incorrect materials and possibly workers not working when they should, although this is very rare in any well run organisation. The planners can also use the measurements as a basis to calculate the labour cost for products and interim products.

Finally, as mentioned, improvements to the processes may be needed and work measurement allows planners or production engineers to identify these. Finally, they are able to evaluate proposed improvements to the processes and later check on their success or otherwise.

Establishing work standards can use a number of available methods. The first is simply to use historical data, which is the easiest method provided the data has been collected systematically and is reliable. It should be safe for use for routine activities which happen for all ships, particularly for parts preparation and the early stages of assembly. Statistical analysis of previous, similar work will give standards then which are useable. This is a cost-effective method, though there is a danger in using such past data if there are any inefficient activities so this must be carefully considered.

Using stopwatch studies is also a possibility in some activities, mainly those with repetitive operations. Here an observer measures the times taken for work to be performed. The method can be used to develop systems for piece work and bonus payments.

Pre-determined time and motion studies use standard times for elements of work and these are then used as building blocks for the total time for a piece of work to be completed. The method is generally applied to detailed assembly operations with short cycle times in mass production. However it is also usable for ship production but at a macro level with times measured in minutes or even hours, rather than seconds.

Work sampling involves taking random samples of activity during a day. The activities to be studied are recorded from time to time during the working day, at random intervals over a period usually of several days. It is applicable to routine, but non-repetitive work of which examples are crane operations or vehicle movements or clerical activities. Over a period the studies can reveal useful information about equipment utilisation as an example.

An extension of work study is discrete event simulation which has a basis in work study and scheduling. This allows a shipyard management, through the production engineering function, to assess the likely effect of work station design and the scheduling of work through that work station. There are several objectives. First to consider how real-life production varies from theory, then to consider the effects of variations and deciding how to manage these. It has been used in some shipyards, but is best applied when the improvements available from conventional work study methods are exhausted.

Conventional analysis and scheduling of production systems makes a number of assumptions. First, that process times are fixed, or subject to regular variations, second that the process times do not change once in progress. Further, using average data, such matters as inter-arrival times are assumed to be consistent or predictable.

In reality, changes can occur in process times, not always predictably. Some mistakes are made in production requiring rework and some materials may arrive early or late. The workers can be inconsistent, becoming tired at times during the day, typically before scheduled breaks. If the data can be collected on the frequency of variations then a simulation model can be constructed.

In summary, work study analyses a production situation to produce the best design of production system. The work study charts are the basis for this, while other techniques including assembly line balancing are used to define the structure of a production system. Scheduling is then used to determine the timings for operation of the system. Finally, simulation can be used to evaluate the system operation, given more realistic variations in timings of activities.

While assembly is generally thought of as steel structures, there are also outfit assemblies. These are also created, in similar stages to steel and include minor assemblies for examples pipes and their supports. The principle is that the earlier work is completed, the fewer worker hours will be required. Also, more parallel stages of work are created which can shorten the overall time to complete a ship.

Pipe and other minor assemblies can be assembled together to create engine room modules. These are complete sections with all equipment, including electrical items, pumps and all systems assembled on to a foundation. Using these a machinery space can be largely outfitted with the minimum need for lifting individual parts and very little manual effort. For large ships, machinery space blocks can be assembled, with significant outfitting on the hull structure.

Some equipment can be purchased as a unit from the supplier, so that it is basically ready to install with minimum input from the shipyard workforce.

Chapter 14
Ship Construction

Nothing is less productive than to make more efficient what should not be done at all
—Peter Drucker

The construction of a ship's hull and the completion of outfitting of that ship, are major operations in the building process for most shipyards. It often requires the largest number of workers of all the activities, depending on the size of the assemblies which are used. The work performed carries a high cost, partly because of access difficulties. As has been described the dock, main cranes and where construction is undercover buildings are the largest structures and most expensive in the shipyard. The dry dock, or other final construction area, is also the critical part of a marine production facility. The speed with which the structural units and outfitting can be installed will largely determine the overall performance of a facility. The faster a ship can be constructed and outfitted, the more ships can be completed and that should be a benefit to the company. So having all work complete before interim products reach the dock is a main goal in production.

In the past ships were built on a slipway adjacent to sufficiently deep water, and many shipyards continue to do so. During construction on a slipway the ship is supported on blocks, which are high enough to allow access for workers beneath the bottom of the ship. The completed ship, or at least the hull once watertight, is then transferred onto "ways". These are usually two parallel beams, running down to the end of the slipway and some distance underwater beyond that point. These fixed ways are built on the slipway, and then sliding ways placed on top with a layer of grease or other low friction substance between the two. The ship is held in place by a series of "triggers" which prevent it moving until the shipyard is ready.

When the triggers are released, the ship will slide down the ways and enter the water. The ways provide support until the ship is able to float off them. There is some risk that the ship may not move, in which case force is applied by hydraulic rams. Once the ship enters the water the stern lifts when there is sufficient buoyancy and the ship begins to float. And at this point the ways may partially collapse, causing damage.

G. Bruce, *Shipbuilding Management*,
https://doi.org/10.1007/978-981-15-8975-1_14

An advantage of a slipway is that the ship is at ground level during construction, making access somewhat easier, and the capital costs are significantly lower than those for a dry dock.

An alternative, initially used for small ships, is to build on a flat area beside the water, with access to a lift which is used to lower the completed ship into the water. Such lifts have been in use since around 1960, and have been available in increasing sizes, with a maximum of around 25,000 tonnes. In the early use of lifts, they were often also used for shiprepair so that ships could be lifted from the water and moved onto a flat area for the necessary work to be carried out. In a busy shiprepair operation, the cost of a lift is similar to, but sometimes higher than, a single dock, but the additional spaces which are then created on the land are much lower in cost. For ship construction the lift is often in a shipyard which carries out both construction and repair.

A variation of the above is to construct the ship on a flat area, then load-out the finished ship onto a floating dock, which can be used to lower the ship into the water. This is usually again associated with a shiprepair operation where the floating dock is already available and it provides a low cost option. Some shipyards use a floating dock routinely for ship launch, others when their existing dock or slipway capacity is already filled with orders.

Many other ingenious methods to provide a construction site for the ship and then move it into the water have been used over the years, but they are usually to overcome some obstacle or to take advantage of existing equipment.

The choice of facility is therefore wide, although the usual choice in a modern shipyard is a dry dock. This is a rectangular structure, of sufficient length and width to contain the largest ship which the shipyard is considering constructing. The depth of the dock allows the ship to float off the support blocks, once the dock is flooded, at the appropriate state of the tide if any. As with a slipway the support blocks allow access under the ship during construction. The dock entrance is sealed by means of a moveable gate, which for a smaller dock may be hinged at the side to allow it to open. The gate may alternatively be hinged along its base so that it drops into a recess to allow the ship to float over it.

The most usual gate is a caisson which is a steel structure not unlike a pontoon. This can be flooded to settle into place to seal the dock. When a ship is ready the water is pumped out to allow the caisson gate to float, so it can be towed away from the entrance to allow the completed ship to be moved out. This type of gate can be reversed after each float out, which allows maintenance to be carried out at any time.

Some docks have been designed to be capable of lengthening, for large ships but also to allow more than one ship in the dock during construction. This is one means of increasing the number of ships built in a given time, by building the stern part of one while a second is being completed. Once the second ship is complete it is floated out of the dock and the part ship can then be moved into the section of dock it has vacated. An intermediate dock gate is used to keep the part ship dry during the process. It is usually moved, once the main dock gate is back in place, on sliding ways, similar to those on a slipway, but horizontal rather than sloping. This approach allows twice as much time for the construction then outfitting of the after part which has the main machinery and the deckhouse. Occasionally longer or wider docks are found, so two complete ships can be constructed at the same time.

Any of the construction areas described may be partially or completely covered for weather protection. It is easier to cover a dry dock than a slipway, because the depth of the dock reduces the necessary building height. Even then the building has a very high cost, and as usual the proposed investment has to be carefully reviewed and is usually only viable for high value ships.

The final construction area is typically the area of highest capital cost, mainly because of the high civil engineering costs of the dock. Unless the dock is built on rock at the ideal depth, it will have to be piled for support and to anchor it to solid ground at depth, or made of mass concrete so that its weight will anchor it. Occasionally docks are continuously pumped to remove water from below to avoid damage from the upward water pressure under the dock floor. Further cost is incurred from the equipment for movement, storage and especially handling large assemblies. The alternatives available for the handling of these assemblies, which include blocks up to around 1000 tonnes and sometimes ship sections up to 5000 tonnes, include level luffing jib cranes, goliath cranes, overhead cranes in buildings, self-elevating transporters, in association with cranes and ground level systems using multiple wheels or sliding or "walking" system.

Many production facilities are designed for very heavy lifts and specialised movements of heavy products, including cases where ships have to be moved from the land onto a floating or into a dry dock. Complete ships of up to 40,000 tonnes lightweight have been moved in this way.

There are various ways of constructing hulls. Going back to the construction of wooden, iron and early steel ships, the hull can be built from small pieces. In the latter half of the twentieth century, with a few exceptions, building the hull in units has developed. The unit is assembled from a number of parts assembled away from the dock, preferably in a workshop. From around 1960, shipyards began to combine units into larger blocks before these were moved to the dock and as ships became larger "grand blocks" were introduced. In a few shipyards, the units are assembled into "rings", that is a length of the full hull cross section.

The general trend has been towards larger blocks. The motivations for changing the hull construction method are both cost and, especially as larger ships began to be built, the time taken to construct the hull. Using large blocks reduces the man-hours to complete the hull, saving costs, and also increases revenue by faster construction. This allows more ships to be built in a given timescale. Better use of the increasingly expensive facilities also assists in reducing the associated overhead burden, as the costs of paying for the dock and cranes can be spread across more ships.

The key objective is to complete as much work as possible at an early stage. This means the work can be carried out undercover, with good access, down hand and with mechanization of some processes. The benefits can be illustrated by considering the ratios of man hours spent in different stages of production. The figures quoted below can be considered to be typical, and others can be found, but the overall range is similar.

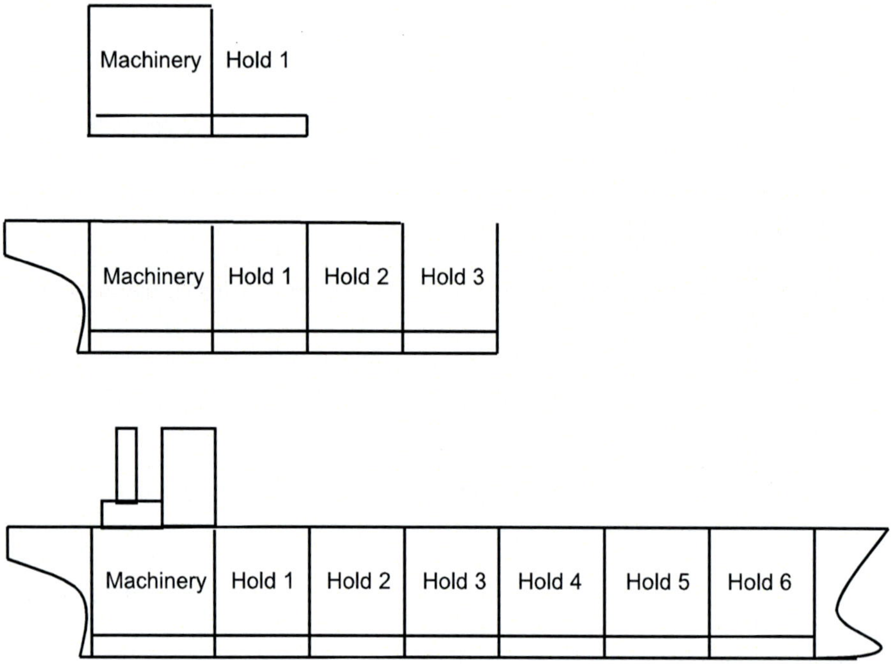

Fig. 14.1 Ship construction typical sequence

For a given piece of work, such as welding two parts together or installing a pipe, the ratios of worker hours required for completion at different stages of the production process if the requirement is one hour at the sub-assembly stage are:

Work stage	Worker hours
Sub-assembly	1.0
Unit assembly	2.0
Block assembly	4.0
Ship construction	8.0
Afloat	12.0

Once the workers and equipment are in position, the actual task may take a similar time, except where overhead work has to be carried out or otherwise in difficult conditions. However the extra time will be needed to move the workers and all the required resources to the work site, which can be relatively inaccessible on the ship. Portable services are required and often some access equipment.

As shown in Fig. 14.1(1), the construction of the ship generally starts at the machinery space. This is so that work in that area is progressed and then the outfitting of what is usually the most complex space on a ship can begin as early as possible, as in Fig. 14.1(2). Most ships also have the accommodation block directly above the machinery, with the crew quarters, services, navigation and other management requirements. Again, once the machinery spaces are competed the accommodation block can progress as shown in Fig. 14.1(3). While the machinery and accommodation are completed the rest of the ship, for a cargo ship relatively straightforward steelwork with limited outfitting, can proceed at the same time.

Until about 1950, ships were built piece by piece. This construction has a very low investment cost, because there is minimum craneage and the items to be moved require only small scale transport. This method of construction can still be found in some places, but it is only appropriate for very low labour costs. This is because all fairing and joining of the parts has to take place in the open so work may be affected by weather. There is also a slow speed of construction and limited opportunity for outfitting until the hull is nearly complete. There is a lot work in progress as the ship construction proceeds slowly, and as a consequence there is potential for steel corrosion before the ship is completed.

Construction in small units is common for small to medium sized ships. Here the individual pieces and sub-assemblies are joined together in a workshop to form small units. A typical unit consists of a panel of steel, made from several plates, which will be part of the hull or a bulkhead. Panels may be flat, for decks, bulkheads and midship side or bottom units, or curved where they are part of the hull towards the fore and after ends of the ship. The panels also have frames, floors, girders and or other stiffening attached. Some outfitting of the units is possible.

The work content at the hull construction site is considerably reduced, and this can shorten the time taken to construct the hull. The man-hours expended are also reduced, since the work completed at an earlier stage requires fewer man-hours. There is additional investment required especially in terms of cranes and other handling equipment. The units have to be accurate, although adjustments can be made when required, but requiring additional time and so increased cost.

However the process is still relatively slow and the labour cost is still high. If good accuracy can be guaranteed then construction may start in two or possibly more locations. Accuracy is more important because the final units then need to be inserted between two already in place. Accuracy of unit assembly will allow larger and more complex units to be produced, further reducing the time and effort at the construction site.

If the shipyard has adequate cranes, and is capable of good dimensional accuracy, then especially for larger ships the best alternative for construction is to use large blocks. This is common for large vessels, and some smaller ones. The term block usually implies a large structure, possibly up to one thousand tonnes for very large ships, and requiring large capacity cranes for construction. However, the term block here is used to mean a structure which is a significant portion of the hull structure. A typical block would be the bottom structure over the length of a cargo hold or one side of the bottom structure with the side structure attached. So for a smaller ship, the block may be modest in size compared to a VLCC for example.

There is a need for highly accurate assembly, because the blocks are rigid in three dimensions, compared to a panel unit which is only rigid in two dimensions. The labour cost is significantly reduced compared to unit construction, because more work can be transferred to an earlier, usually undercover stage of assembly. This reduces the work content at final construction.

There is a larger investment cost in facilities, in particular cranes and equipment for handling and moving the larger structures, as well as the associated buildings. The work has to be completed in a particular sequence, so there is less flexibility compared to unit construction. There is also very high reliance on management control and on accuracy of the structures taken to the construction area.

Several blocks can be combined to create a ring structure, basically a transverse slice through the hull. This is not a commonly used method, and it applies generally to mid-sized vessels. The ship is constructed from what may also be called "super blocks" or "grand blocks" which are a complete hull cross section, the length of the largest steel plate which can be handled in the shipyard. The ship construction process can be carried out substantially undercover. The method was adopted in the 1960s and 1970s in a number of European shipyards, where the rings were completed in a building and then slid out into a building dock, or onto a slipway. Other shipyards used an elevator to lower the rings into a dock. Some of these shipyards are still operational. Using this approach, the labour cost is further reduced and the cost of the construction building is lower.

The rings can be assembled from units which gives a degree of flexibility, or from blocks. However, this approach is totally inflexible in terms of sequence of work. The after end has to be installed first, then machinery spaces and subsequently the cargo carrying portion of the hull. This lack of flexibility in sequence of work means that the planning of operations, supply and installation has to be certain. As with the use of large blocks, the size of the rings is relative to the size of the ship, so they are not necessarily very heavy.

Where the rings are used they will usually be assembled in a workshop at a distance from the construction site, and then moved by multi-wheeled, self elevating transporter. It is then sometimes possible to use the transporter to position the rings on the slipway, or level building site. This avoids the need for vary high capacity cranes. Some ships have been built using ring structures from several shipyards, with final ship construction at one of them. Such rings can be moved on barges, then slid into position for the construction.

The decision on which hull construction method to use is based on a detailed analysis of the production requirements. The planned construction time for the ship is the starting point because speedier construction will require larger assemblies to be used. An initial breakdown into units and stages of assembly is made for each of the alternative methods being considered. Then the work content for the ship is estimated, and used to determine the worker hours. Knowing the reduced worker hours associated with completion earlier on the assembly process, and so minimizing the work at the construction site, will allow the likely cost to be estimated for the alternatives.

For each of the potential construction methods, the cost of investment in cranes, large buildings, heavy movement equipment and other items is determined. The additional investment required so that the larger blocks or assemblies can be used is then compared with the reduction in labour costs, for the number of ships it is expected to construct over a future period of time.

It is also essential to consider the feasibility of all the alternatives. There may be questions about the accuracy of assembly, which can lead to rework for example. The best solution can then be selected for investment.

Coatings are assuming an increasing importance in marine operations. Modern underwater paint coatings are designed both to provide protection from corrosion in hostile environments and to have a long life to minimise the requirements to dock ships for maintenance. For many ships, internal coatings are also important. This can be to protect the cargo spaces from chemical and other hazardous cargo. It is also necessary to maintain the condition of the steel in the separate ballast tanks.

The application of coatings in ship production is assuming much greater importance in parallel with the operational importance and the increasing cost of the material and the coating application. Historically, steelwork was regarded as the most important element in ship construction. More recently, outfitting has been seen to be of at least equal importance in achieving good ship construction performance.

Current thinking gives coating a similar importance, because this can be a bottleneck process in shipbuilding for the most advanced shipyards. The development of better production technology has resulted in parallel outfitting and steelwork, and much reduced final construction times. Assembling very large blocks, and outfitting them extensively, gives very short times at the final construction area.

To allow very short construction times, steelwork has been extensively automated. Outfitting is more difficult to automate, but progress has been made with pipes in particular. Work on board has been reduced by use of outfit assemblies and installation of outfitting onto blocks. Cleaning and coating remains a labour intensive operation so it can become a serious bottleneck. Even where large paint coating cells have been introduced for the final coating of large blocks and units, the labour intensive nature of the work makes it difficult to reduce the time taken. The required curing time for recently developed coatings also limits the ability to reduce the overall coating time.

Owners are increasingly seeking to reduce ship operating costs and minimising the need to dry dock ships is part of this. The coatings can help to reduce fuel consumption, provided they are carefully applied in controlled conditions. The coating manufacturers who are offering the long life coatings demand their application in good conditions, and shipowners will dictate these in order to have a guarantee of the coating performance. The shipyard is left to manage the effects on the overall shipbuilding programme.

Increasingly, the management of the coating programme is carried out by the shipowner's superintendent and the paint company representative. The contract is often between the owner and the coating company for the coating supply, and this can cause some of the problems of owner supply discussed in Chap. 8. The paint company will then determine the coating scheme and the operating conditions for application. In these cases, the shipbuilding company simply follows instructions, which relieves them of responsibility for any problems associated with application in unsuitable conditions. However, they still have to follow the coating scheme instructions, and if there are consequential delays to the programme this will be a potential source of friction between them and the owner.

Coating companies will now provide a detailed specification for the work, including details of the coatings, application areas, procedures to be followed, surface conditions required and sequence of work. This will include the curing requirements prior to over-coating, which sets the criteria for paint cells and to some extent dictates how long a block takes to prepare.

Coating has to be done after all other activities, in particular all hot work, have been completed in order to avoid re-work. It has been estimated that in some shipyards 25% of coating hours may be required for rework if the work is not well co-ordinated with other activities. For some ships, coating may be 5–10% of the total ship cost, so rework can be expensive. Apart from extra costs to the shipyard, the quality of a repaired coating may be compromised.

Many shipyards try to install as much outfitting and then as much of the coating before the block assembly stage. This reduces the man-hours required, and leaves only limited areas of the steelwork requiring final coating. The problem is that where coating is achieved at an early stage of steelwork, all the succeeding stages of the production cycle are potential causes of damage.

Causes of coating rework are many, with mechanical damage caused by transport, lifting assemblies onto supports, fairing and alignment operations and installation of equipment. Hot work causes damage, from tack welding for foundations and supports, repair of poor welding. Nearby shotblasting can damage coatings, as can simply storing coated steel in the open air if there are any weaknesses in the coating.

The preferable solution is to avoid problems by addressing the design of the ship to facilitate application, although the extra casts incurred must be outweighed by the savings. It is also important to use the most appropriate facilities and equipment to avoid damage, improving the planning and organisation of the work to avoid late outfitting. There are operational solutions to unavoidable problems, such as using temporary protection during other operations. Reducing the amount of hot work and other improved working procedures will assist. Importantly, attention must be given to the maintenance of all procedures once they are established.

The planning and organisation of work must be amended. Typically shipyards begin to develop a build strategy based on the requirements of steelwork production. Most have now also recognised the importance of also taking outfitting strategies into the development of the planning of the strategy. It is increasingly necessary to also consider coating requirements in the strategy and the subsequent planning and scheduling. A cleaning and coating programme is clearly essential for scheduling the paint cells, but the work at other stages of production must also be planned alongside other activities.

Labour cost control in coating is required, and can be achieved by monitoring all activities to eliminate rework, so that coating is only done when other work is finished. The labour force in general must be adequately trained in the importance of preservation of coatings once they have been applied.

Quality assurance has a role, in several aspects. Chief of these are the setting of finished quality standards in agreement with the ship specification and procedures which will assist the stability of the production processes.

The technological environment is the one with which most shipbuilding managements are familiar. Much of this book concerns technology developments in the shipbuilding industry, and the use of the technology in modern shipyards. Most developments have had the intention of greater productive efficiency in the shipyard and also creating an improved product. As an example joining the steel plates from which a ship is constructed has moved from rivets to welding, reducing the man-hours required and reducing the weight of the ship at the same time because plates no longer overlap at the joints. The weld is inherently stronger and watertight.

Early welding produced many problems, such as brittle fracture of welded ships, and there were many weld failures. It was common for perceived key areas of the ship, such as the bilge plates and sheerstrake, to continue to be riveted as a precaution. However there has been steady development of welding technology especially in the last fifty or so years, and improvements have been adopted by shipbuilders over the same period. Improvements in both weld quality and speed of production have been made.

Developments in welding technology are easily monitored by a shipyard. There are national and international institutions researching and publishing on the subject, and the many manufacturers of welding equipment are eager to sell more products. Most developments take a long time, often decades to move from a laboratory principle to a useable piece of equipment for a shipyard. However, once a new welding process is available for practical use, it is almost certain that suppliers will be advertising and promoting it to the industry in general.

As well as improvements to ships and shipbuilding, the newer processes may be safer, less polluting, quieter, and so more acceptable in the increasingly safety and environmentally conscious world.

Any shipyard management needs to maintain reasonably close contact with developments in the key processes in use. Apart from publications and manufacturer information, visits to other shipyards which are using new technologies may be arranged. Some large shipyards carry out their own, in-house research on both ship product development and building process development. Collaboration with other shipyards, often on a national level, and with research and manufacturing organisations is also common. Governments will sometimes support collaborative research and development projects.

Many technologies proposed for the shipbuilding industry turn out not to be practical, or are too expensive. They may be too slow for the heavy materials used, or require very high levels of accuracy, or protected environments. This is no reason not to keep up a continual review of what may become available.

Some processes may only be useful only for niche areas, for example laser cutting and welding. Others may be promising, but the time from demonstrable to efficient in production for a new technology can easily be twenty years. A look back at welding, plasma cutting and particularly information technology will show this.

Some very promising developments have proved difficult to use in mainstream ship construction. Aluminium should have uses in superstructures in particular, but has only been seen in niche applications, such as passenger ships and small, lightweight vessels. Fibre reinforced plastics (FRP) are good for small vessels, but have been limited in use elsewhere. Another development which has been struggling to find a mainstream role for twenty years or longer is the sandwich panel. Two thin steel plates with a lightweight core, currently a resin is favoured, should offer an ideal way to build deckhouses. Problems of forming any shaped structures and the joining of panels makes their use in hulls and main structures difficult. As a result the application is mainly used for repairs to steelwork in ships in service.

It is important for a shipyard management to monitor technological developments, but their use depends on suppliers willing to invest and produce products at acceptable cost. It also depends on acceptance by regulators who are understandably relatively conservative. Being first with a new development can be a dangerous move, in terms of the basic objective of business to make money.

Chapter 15
Quality Management

Quality means doing it right when no one is looking
—Henry Ford

Quality management is an important part of industry in general, and formal quality management is mandatory for many companies in that if they do not have an acceptable, formal quality system customers will not consider their services. In practice this requires a company to have created a quality management procedure and had this accredited to an appropriate standard, typically set by the International Standards Organisation, discussed later in this chapter. When it is formally defined, quality management requires quality planning, quality assurance, quality control and continuous improvement. The overall stated objective of these is to ensure that the production process is so organised that it will ensure that the finished products satisfy customer requirements.

Quality may then be defined as "fitness for purpose", which means that the product must be "good enough" for the customer's requirements. Another way to consider this is value for money, so that the term quality does not necessarily imply an especially high standard of product.

The first requirement is quality planning. This will describe the system of quality management to be developed and the measures which will demonstrate that quality has been achieved. The system has to be agreed with project stakeholders, and for some projects this can be complex, especially for a novel product.

The quality of a ship is largely defined by the initial specification, and if the finished product matches the specification, then the quality has been achieved. But to reach the required quality, the production process has to be carefully managed. In shipbuilding, there are internationally recognised requirements, and national regulators, so ship construction will necessarily have to take place in a well organised structure. These are largely concerned with the safety of the finished ship, and on top of these are more specific owner requirements, for production standards. These are also manageable through recognised systems for example for dimensional accuracy.

The quality plan for a shipyard also has to consider the efficiency of the ship production process. It is very easy to build a ship to conform to a standard if errors

G. Bruce, *Shipbuilding Management*,
https://doi.org/10.1007/978-981-15-8975-1_15

are made and so rework is then needed, but it will not be built without at least some losses. So a plan is required, followed by a quality assurance programme which is set up to ensure that procedures are followed and that staff are correctly trained. The quality assurance function for a shipyard must be independent of the individual projects in checking that procedures are followed and also in the inspections which are used to check what is actually happening to the interim products.

Once a ship project is in progress, quality control is the means of ensuring that the interim and final products conform to specification and are acceptable to stakeholders. Quality control is the inspection and testing of interim products, so the necessary activities will check those products against the acceptance criteria. It is possible that, once agreed, the specification may need to be modified. Commonly this is to accommodate change requests, while maintaining acceptable time and cost constraints. Any consequent changes to acceptance criteria should be approved and communicated to stakeholders.

After completion of a project, as has been emphasised, some evaluation is essential to look for good practice to adopt and to learn lessons. As part of the quality management, this is generally described as continuous improvement. Some improvement can be made during the progress of a project, in the more routine activities which are part of it. Information from previous projects should be consulted at the beginning of every project, and ideally any useful lessons should already be incorporated in the current quality procedures. This ensures that any relevant information is used in the preparation of the new project so providing the continuous improvement which is an important management goal.

It matters that all aspects of quality management are in place in a company. Without the plan there will be no procedures set up to be followed. Then the procedures should lead to accreditation, which is the award of a recognised certificate of quality assurance by a qualified authority. Without this it is very difficult to obtain any orders. Quality control, checking on what has been done is the traditional application of quality. This must be done, even where there are good procedures, because the overall objective is that the product reaches the necessary quality standards. Finally, continuous improvement follows from the rest of the quality process. If any of these are missing, or incorrectly carried out, then the quality management can become a set of empty phrases.

There are different views of quality. For example the finish in crew cabins on a cargo ship would not be acceptable to a passenger in a premium cabin on a cruise ship. And the requirements for a luxury private yacht will be much higher again. A quality product can be stated to be one which is acceptable to the customer, but the product must also be produced at an acceptable cost to the producer. As a result, achieving quality is very much about managing production as well as product quality.

There are several aspects of quality. Product quality determines the acceptability to the customer, starting with a suitable specification. This will confirm that the ship is of adequate design. This will be in terms of functionality so that as long as the ship performance is as specified, and the ship performs reliably, the owner has a quality ship. The ship should also continue to operate in a satisfactory manner, so durability is part of quality and the ship will last for an acceptable time. Then a further issue is safety, so that the crew, the cargo, in some cases passengers and the marine environment will be protected from any dangers.

The production of the ship must ensure that when complete there is conformance to specification. Consistency is also important, so that all elements of the ship are acceptable to the customer. However from the viewpoint of the shipyard which is constructing a ship, the specification must be achieved without excessive effort, or any rework of elements which do not conform to the specification at the first attempt. It is relatively easy to achieve the specification, if a lot of time and effort is used, but this can make the production cost unacceptable. The definition of quality which is in the specification must be followed closely, to avoid an owner or their representative trying to obtain an outcome which is beyond that specification. Sometimes arguments about the quality are used to delay acceptance of a ship.

In essence there are two ways in which adequate quality can be achieved. The first, long-established way is to inspect work which has been completed to check it conforms to the appropriate specification. This has some important applications and is a very valuable method in the right circumstances. The second way is to check the quality at stages during the process, then use the results to create a database. This database can then be used as a starting point to re-design the production processes so that the required quality is built into the work. This should guarantee quality in the interim and final products. The two approaches can be described as quality control and quality assurance.

Quality control is based on inspection which produces data. This data can be used to construct various visualisations, including control charts, flowcharts and cause and effect diagrams, all of which help to understand the quality of work and how it may be improved There are a number of ways of reviewing the quality of an operation. Acceptance sampling is concerned with reviewing the acceptability of the inputs to and outputs from a process. Acceptance sampling is the oldest technique and at its simplest, acceptance sampling is carried out for all products. For a ship, if the required speed is not achieved during the sea trials, then it may be rejected. The contract will usually include other requirements where failure may lead to rejection of the ship. It will be clear that to allow a situation of rejection to develop is disastrous for the producer, and potentially equally so for the customer. Even if the ship is not rejected, there will be a serious cost penalty for the shipyard. Avoiding any of this is an important part of the quality management process.

Testing every item produced is described as 100% sampling, This is not usual practice, except for large made-to-order products, such as ships where there is only one product or perhaps possibly a small number of products and it is necessary to be certain the quality is correct. Full sampling may also be required for critical items in any product for example where there are major safety issues, or for military use.

However for many products made in large numbers, and for bought in parts or interim products in ship construction, 100% inspection is typically not feasible or is just uneconomic. Usually therefore a sample of a batch of products is tested. If the sample is found to be acceptable, then the batch is accepted. If it is not acceptable then the whole batch can be rejected. This is a sensible approach for many businesses, where large numbers of products are made and where the customer can quickly obtain

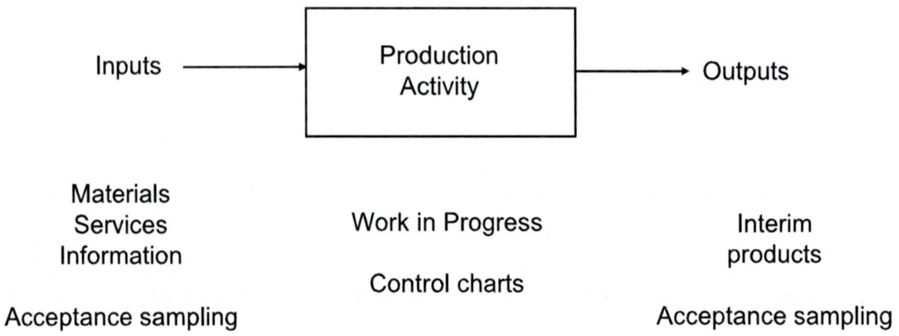

Fig. 15.1 Activity model

a replacement batch of a supplier's products, or can readily switch to an alternative supplier. Supermarkets are a prime example of such a business.

Obtaining replacements is often not an option for much of the material supplied to a producer of made to order products, such as ships, so sampling has to do more. Because only a sample is tested, there is therefore a small probability that batches with defective items will be accepted, although robust sampling methods should minimise this potential. This represents a danger for the shipyard customer. However, there is also a possibility that although the failure of an inspected item results in rejection of the batch, the other items from that batch may be acceptable. This is a danger for the producer of the sampled items.

In contrast process control is about managing the process to ensure that acceptable inputs and outputs will be obtained because the process itself is under control and will provide correct results because of this. The basic technique is the use of control charts. The activity model in Fig. 15.1 illustrates the use of acceptance sampling and control charts.

Quality management is of importance through the whole scope of a ship construction project. In the early stages of product design, it is about correct design and specification, leading to adequate materials, equipment and then systems which will ensure that the ship will be suitable for its intended operations. During the design of the production processes, quality management is to ensure the ability to deliver the product design requirements. Planning of operations and scheduling of work, including the storage of materials is monitored for adherence to procedures.

The materials for the ship are managed to ensure there is conformance to specification and suitability for the operations. During production of the ship, the quality function will monitor the operation of the various processes to ensure that procedures are followed and the resulting work is correct. A part of this is to ensure that the employees have been given suitable training, motivation and that the organisation of their work is appropriate.

Inspection is important and varies for different interim products and stages of a project. At the end of the project, the ship is trialled at sea, with all important operational aspects checked against specification. This is clearly not a good time to find any problems, so earlier inspections also have to be carried out. The main engine will be inspected during its production by the manufacturer, and prior to delivery will be run on a test bed so it is seen to be acceptable for both the shipyard and the shipowner.

The operating systems of the ship will be tested in use during trials, but again earlier testing is important so testing of partially complete systems will check for defects during production and installation. Some testing, usually sampling, will be carried out for system components such as pipes. Installed equipment, for example a winch will be accepted from a reliable supplier, with limited inspection for possible damage on delivery to the shipyard. Welding will be sampled, except for any critical items where one hundred per cent inspection may be mandated. Where automated equipment for welding is in use, the welding process will be inspected on successful commissioning and then welds will be accepted, subject to the correct maintenance being carried out and some occasional checks.

In order to be reasonably certain of consistent outputs from a process, the process itself must be stable. In a ship construction environment, stability is reached in various ways. Ideally it is reached by repeatedly making similar products in the same work station, with the same people using the same procedures and the same equipment and processes. There will require to be some monitoring of the inputs, the process itself and the outputs from the process.

Stability can be measured using the results of the monitoring. The outputs from a process, the interim products, will be checked for specific characteristics, such as dimensions, strength, shape, finish and so on. The sample outputs generally follow a normal distribution if the process is stable. That is there will inevitably be small variations in the outputs which are acceptable if they fall within some specified limits.

Prior to industrial development, human artefacts were made by craftsmen, individually or in small groups, including ships. Although the shape of the ship hull was sometimes defined by a half block model and faired lines, the final ship only approximated to that shape after its construction. Parts were adjusted to fit as the ship was constructed. The development of industry, early mass production and particularly interchangeable parts initially for armaments, led to a need for dimensional accuracy. Shipbuilding accuracy was less critical, with overlapping steel plates riveted together and piece by piece construction. This allowed parts to be manoeuvered into the correct location with adjustments as they were joined together. Once ship units, then blocks were used in construction, accuracy became critical. In fact the tolerances in shipbuilding given the scale of the final product are actually similar to those industries making far smaller items.

As accuracy in industry increased, a need arose for inspection to ensure this was maintained. However the limitations of simply inspecting work after completion have been discussed, and it was realised that developing a process that ensured accuracy was required.

For industrial products, tolerances are set within which small deviations are acceptable. Acceptance sampling will detect out of tolerance items, which will not be useable. Although acceptance sampling will avoid either defective inputs to a process, or defective products reaching a later process, they cannot assist in avoiding

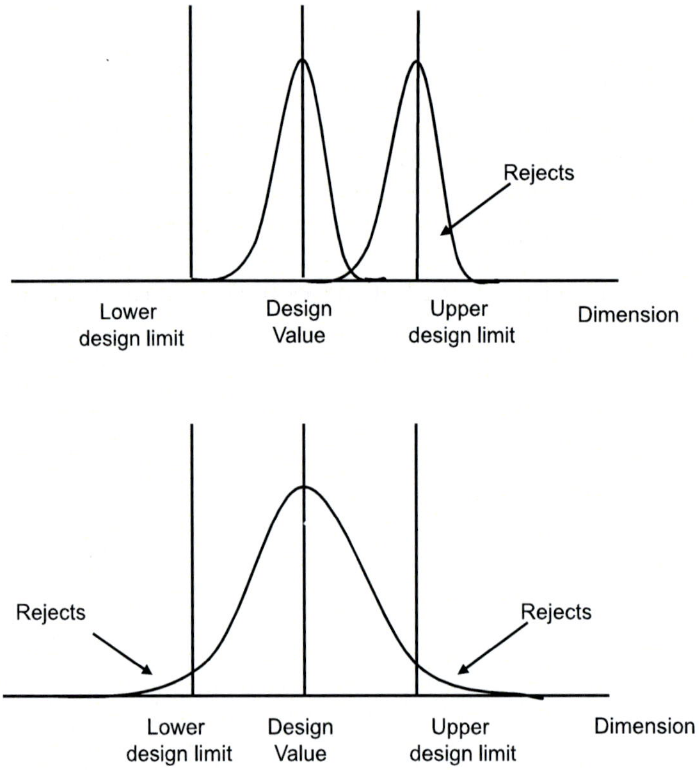

Fig. 15.2 Dimensional tolerances

production of defectives in the first place. In particular for large, made-to-order products, out of tolerance items can disrupt production schedules. If an interim product is too long or short, and the fitting together of two interim products depends on the correct length, then the fitting process will be delayed while adjustments are made. As the ship on completion depends on a string of products joined together, then it is also easy for inaccuracies to get out of control and correct overall dimensions will not then be maintained. The result will be extra work to make corrections, leading to losses.

Figure 15.2 shows the use of tolerances. In the upper diagram, the distribution shows how the dimension to be measured may vary around the design value. In the first case the mean value of the distribution is coincident with the design. The values in the distribution also fall within tolerance limits. These limits determine whether or not the part produced is acceptable, so if it is within the limits then it is going to fit in the location it is intended for. These limits may refer to an absolute size, or for example to the gap between two steel parts suitable for welding.

The second part of the diagram shows the same distribution, but here the mean value is coincident with the upper tolerance limit. In this case the process producing

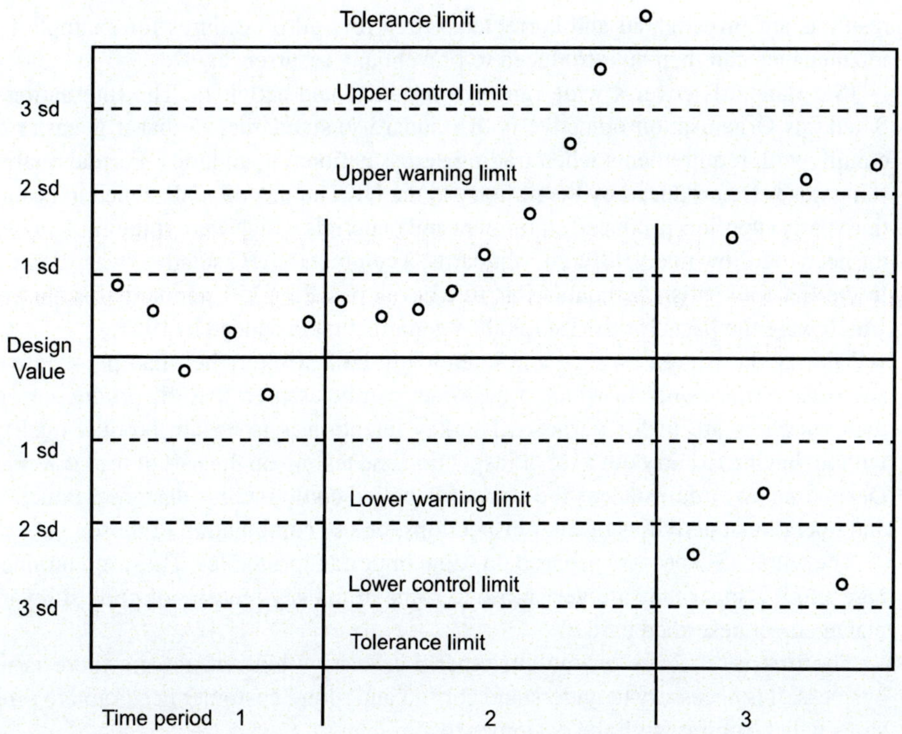

Fig. 15.3 Typical control chart

the parts is giving results which are permanently too large, and it can be seen that 50% of the parts made will be too large to be used.

The lower diagram shows different results from the same process. Here the parts still have a mean value which is acceptable, but a proportion of them are too large or too small, because they are beyond the tolerances. Such parts would not fit due to overlapping, or would result in a large gap between adjacent parts, so for example welding would be a problem.

It is important that situations as presented here should not be allowed to happen, so a means of predicting and then avoiding them is needed.

Figure 15.3 shows a control chart, with the tolerances and the warning and control limits. In the first period, the process displayed is under good control because all the measurements are within one standard deviation of the mean. In the second time period, the process is rapidly going out of control and the final measurement is outside the tolerance which has been set. By reviewing the control chart before that point is reached, the trend can be identified and action taken to identify the cause of that trend before it leads to unusable outputs.

The third time period shows the process to be under control but with a wide range of measurements. This also suggests there may be an underlying problem and would therefore also be investigated to improve the process and return to the situation in the first time period. The critical aspect of this is that all deviations from expected result as are investigated and corrected. Often revised procedures for example for maintenance can then be introduced to prevent any recurrence.

ISO standards cover a wide range of products and activities. The International Standards Organisation objective is to standardise standards, so that it is easier to comply with requirements when trading across national boundaries. Formal quality standards were produced by the military in the UK and the USA, designed to ensure third party suppliers produced equipment and materials which were suitable. To avoid the need to follow many different standards, a common NATO standard was adopted. In the UK the British Standards Institute developed BS 5750 standards for quality, which were the basis for the ISO 9000 standards first produced in 1987.

The standards are designed and intended to be flexible rather than prescriptive, and offer a framework in which a customer can be assured that the products from their suppliers are fit for purpose. The key intention is to ensure product quality, without having to carry out a lot of inspection and testing on the part of the customer. Depending on circumstances the standard applied could include design, production and operational activities or final inspections, or any combination of these.

The whole ISO process is based on some important principles. These are outlined here and it is clear they are very close to some of the key features of good shipyard management described earlier.

The first is to focus on the customer, because without customers there is no business. It is necessary to understand current and future customer needs and to offer ships which comply with the customer requirements.

Leadership has been highlighted both for the company management and the individual ship project management. Both have to establish a common set of objectives and then create teams which are fully engaged with these.

Since it is certain that people build ships, and everything else is there to support them, they must all be fully engaged. It is also certain that the knowledge which has been gained by the people of a company far exceeds that of the small management team. By gaining the full involvement of all workers, their capabilities will be available to benefit the company.

The company must be treated as a series of activities which can be improved, and so concentration of management must be on these. They can be considered as a set of processes, with inputs and outputs to be managed, which can be amended if necessary.

The issue of remaining competitive requires improvement of the company's overall performance. This should be a continuing objective. All decisions on management of the activities need to be based on evidence, derived from data on the activities. Effective decisions are based on the analysis of information gained from production data.

Managing the relationship of the company with all stakeholders is critical. Customers, suppliers, sub-contractors and others depend on each other and in most cases a relationship which benefits both parties to a transaction gives the best result.

It is clear that there is a close correlation between the principles of good management set out in this book and the quality process. So if a company is well organised and managed, then quality should be embedded in the company.

In order to confirm that a company is complying with adequate quality procedures, it must be reviewed by a recognised quality audit organisation, generally accredited by a national government. For the marine industries, the classification societies usually fulfil this purpose. Initially the company to be audited was required to produce a quality manual, containing the agreed detailed procedures. This is no longer required, but there must be quality objectives, a policy and a scope describing what the quality management system includes.

Once a company is awarded a quality certificate, there is a requirement for external audits at regular intervals to ensure compliance. Records of quality management must be kept for inspection to show the standard is being maintained. The audit seeks to identify that the company is operating an effective quality management process.

There are also internal audits by the company to ensure continuing adherence to the quality procedures. This is to avoid what was often a short term effort within a company prior to the next audit. It is still possible to introduce ISO system on this basis, so as to obtain a certificate on the wall but then to ignore any issues uncovered in the audits. To avoid this there is therefore a need for serious commitment to quality management by the senior executives of a company.

ISO is flexible, so the quality system can and should be in step with the company operational processes. This will ensure that those processes are correct for good quality and that the compliance with ISO standards supports efficient operations. It is of no value to adopt an existing quality model and try to force the company to adapt, or worse simply to ignore the standard once the certificate is awarded.

The ISO system should help ensure that the customers are happy with what they receive. If quality is defined as "good enough" for the customer needs, and this is contractually acceptable, then that is what is needed. Fixing occasional problems promptly will appeal to most customers and may well be appreciated as much as never having problems, and is almost certainly less expensive for the supplier company.

Those carrying out audits for ISO compliance may simply inspect and report non-compliances. However they will be more effective if they are able to demonstrate how a process is failing and what might be changed to improve it. If the management of the audited company are not clear where the problem is, they may seek to ignore it.

If a company is seeking to improve its overall performance, then ISO standards can assist when implemented sensibly. Given the need to have the certification in order to work for many potential customers makes this almost mandatory, in which case it is sensible to try to make the whole process useful. This will avoid a common criticism of ISO implementation that it is expensive, time-consuming and creates paperwork.

It is believed that having good quality procedures is a benefit to a company, ultimately in money terms. Arguably, if a company adopts effective management of its activities, it is taking a quality approach already. It is true to say that ISO principles can be implemented for the benefits without the certification, if this is not mandated by customers.

A danger is that the whole process will simply create procedures and then check they are being followed, but there is no mechanism to ensure that the procedures are useful, or even correct. Also, the whole process is not designed to deal with innovations, so for any prototypes ISO offers no guarantees to the customer. This takes us straight back to the risks inherent in novel projects.

In conclusion it is necessary to gain ISO certification if the customers require it, so it should be made useful. It is essential to create and follow good procedures in any case to offer the customer quality and to be efficient in production. Finally do not to lose sight of the close correlation between procedures and good management.

Chapter 16
Human Resources

Management is nothing more than motivating other people
—Lee Iacocca

To repeat an observation which is really obvious, all ships are built by the people in the shipyards. So dealing with the people, the human resources, is a really significant element in shipbuilding management. Whereas equipment can be purchased at any time, workers have to be recruited, and often trained. They generally have a choice of where to work, and so need an incentive to be retained because they are always free to leave for alternative employment. When the shipbuilding industry is busy, there is often a shortage of skilled labour. When there are few orders, shipyards have to reduce their workforce, and those who are released may go to other countries, may find new employment in another industry or if older may simply retire. Then when orders for ships increase the workers may no longer be available.

Marine production generally requires large, complex organisations with a high degree of labour specialisation. Recruiting and retaining sufficient workers with the right skills is a universal problem for shipbuilders. Skilled labour must be retained to maintain the business as an operating concern.

The context in which a company needs to develop people will vary. There are some relatively new companies, usually set up in areas of low wages and high unemployment where there is a strong intention to create industries as part of a regional or national development programme. There is a large investment and such developments are often government supported. In these cases there is a clear and obvious need to recruit and train a labour force. The recruitment is mainly young workers, usually under thirty on average. Such developments are often on green field sites in areas which have not previously been industrial.

Developed companies have typically been in business for twenty years or so. The initial trainees have developed into mid career, skilled workers and if there is limited loss of these to other areas or companies, then training and recruitment can be less in demand. If the industrialisation of the area where the company is located is progressing, then loss of skilled workers and recruitment problems may be occurring. The workforce average age is typically mid-thirties to mid-forties.

G. Bruce, *Shipbuilding Management*,
https://doi.org/10.1007/978-981-15-8975-1_16

Mature companies, usually established for decades may find some of the current workforce is approaching retirement age. These are generally in well developed countries with a diverse range of industries, including newer sectors which are generally more attractive to younger workers. In these cases there is an ageing workforce and problems in recruitment and retention. One popular solution is to try to recruit labour from areas where the shipbuilding is newer, wages are generally lower and a move can be attractive to the workers who have been trained. Typically the workforce average age is mid-forties to mid-fifties.

Depending on the company context, the needs and responses in maintaining a skilled workforce will vary. As a starting point for determining the human resource needs, it is necessary to establish a profile of the business under consideration. This will take into account several factors.

The location of the company, essentially whether it is in a developed area with a large population or is new will be a major determinant of the ability to recruit. Also important is the maturity of the company and the industrial landscape. Unless the company is relatively new, then there will be an on-going need to plan for the replacement of current staff, as they move on or retire.

The product mix is a factor, since basically the more complex the ships to be built, there will be a larger range of skills and higher skill levels required of the workers. Also the technology in use will determine the nature of the skills and the numbers of workers required. The location of the company, whether in an industrially mature or newly developing area will be a factor in how a company responds.

The current intake of apprentices and other trainees, skilled workers and graduates or otherwise higher qualified people will help identify the future requirements. The current employment level of full-time and casual workers and the age profile also help define recruitment needs.

The company profile will be arranged according to the various job titles, categories of skills and succession planning. The qualifications for the entire workforce will need to be established as a starting point for in service training.

Having profiled the existing situation, the need is then to develop an understanding of the requirements to deliver the output that the company wants. There is a also a need to create a training and development plan to meet the future needs of the business, whether by direct recruitment of qualified people, training or a mixture of both.

To determine what needs to be done in a company to ensure sufficient skilled people for the future, the starting point is the future company strategy. This will have identified the products to be made, in numbers and types. This in turn will allow estimates to be made of the organisation and production requirements from each activity in the company. The current productivity will be the basis for future changes, including any plans to invest in new technology. The location of the shipyard in relation to population centres and sources of skilled workers and potential recruits will also be a factor.

From the analysis, a picture of the current and expected future employment in the company will be created. This will be based on the activity map, so that those activities which are most in need of people can be highlighted. As part of the process an age profile of the workforce will be created. This will look usually at 5 or 10 year intervals and identify the activities with highest need. Sometimes particular activities have an older existing workforce, or may have a need for more workers.

The ratios of full time to part-time, casual and sub-contract labour will be identified, and over a period of time trends can be monitored and decisions made. The use of casual labour will depend on availability, on the skills required and when those skills are required. An important consideration is that the formation for a skilled worker takes time. It is possible to develop limited skills quickly, but a fully skilled worker takes several years. Several years experience is also needed before workers are fully competent and can operate without detailed supervision.

Training existing workers is also a possibility, to expand their skills or to train them for a different role as the product mix changes. Overseas workers can provide a solution to shortages. This is in fact a general problem in the world of shipbuilding, with many countries using workers from neighbouring states. This usually occurs when the area with the shortage can offer better payment rates than the workers' home countries. It can sometimes lead to a chain reaction, where several countries are using their different neighbours' labour forces.

Using the activity map for the company, the numbers in each function are found, typically initially divided into groups. In the first instance, these can be simplified to five which are:

Skilled workers who have a specific capability.
Unskilled workers who are labourers or semi skilled who have a limited range of capabilities.
Supervisors, who are very experienced and often well educated, typically required in a ratio of one to every ten workers.
Technicians with specific skills in support of design and production, including designers, welding technicians, quality inspectors and others.
Management who all have specific roles in overseeing the activities of the shipyard.

The workforce has then been classified by age in five or ten year bands, by skill category and in the various activities needed. With this information available, attention can be turned to the future operations of the shipyard.

Once an overall picture of the available labour force has been developed, the capacity of the shipyard has to be examined in more detail to determine the capacity for the various activities. The activity map is a useful starting point again.

Matching the operational capacity of a production system to demand can be difficult, for a number of reasons. This is the case both in the short term for a contract and the longer term when planning the future shipyard strategy. Beyond the immediate order book for a shipyard, the future demand is uncertain and hence the resource requirements are uncertain. It is always possible that the types of ships in the future order book may change and then the balance between different worker categories will also require to be altered.

Large ships may be in demand and the absolute numbers of workers may need to increase. The numbers required may also decrease with a weaker market, and the reduction in numbers of workers employed will have to be carefully managed so as to retain the more skilled and experienced. This is a problem because if a potential reduction in workforce is well known to the work force, it is precisely those better skilled workers who will find it easier to move to other employers.

There may also be changes in technology and substantial changes planned in the facilities and equipment to be used in the shipyard. For an established shipyard these changes will always be associated with a plan to increase production or to reduce production costs for a consistent level of output. Reductions in worker numbers may again be required, or re-training of the existing workforce, again potentially destabilising the workers and leading to unplanned losses of skilled people.

For the short term, a specific contract or contracts may require short term changes in the workforce. So a temporary peak in numbers for a specific category of worker may be needed. It must be decided how best this can be done. Additional hours of work may be used although this would usually be for a very short period of time to be acceptable to the workforce. Casual labour might be employed, assuming that adequately skilled people can be found for the relatively short term. More usually, some of the work will be sub-contracted.

Transaction costs are a potential problem when sub-contractors are used. These are the actual and hidden costs of using those sub-contractors, examples including insurance, contract costs, risk of failure to deliver and quality assurance. These can apply to any supplier to a shipyard, not only sub-contractors, but where a new sub-contractor is used or the decision is taken at short notice there will often be no history with the contractor on which to judge the risk. Or the risk may be accepted because of the urgency of a particular situation on a contract.

Measuring the capacity of a shipyard is problematic in ship production, because of the variety of operations undertaken in the building of a ship and because the future resource requirements are only an initial estimate when the requirements are determined.

Managing the capacity in terms of the labour force is important. It is relatively straightforward to determine average demand levels over the future building programme. The average can be calculated using productivity figures from the past, modified by planned changes in technology and products. This information can then be used to determine the required capacity. However, the demand for resources as set by an unmodified programme will not be consistent, but will have peaks and troughs. The actual demand will fluctuate and it is then necessary to deal with the variations.

Figure 16.1 shows the employment numbers for a company over the duration of a project, which rises to a maximum level then declines towards the completion. If several projects are in progress then the overall variation can be reduced and work scheduling can also help reduce the variation. What is left must be managed. If the labour force numbers are set for the maximum demand, then there is wasted capacity at the start and end of the project, shown by the areas between the curve and the maximum level. If the capacity is set lower, then there is insufficient capacity and additional workers need to be found, represented by the area under the curve. A compromise level of workforce can be set to try and minimise the variation. This would try to set the numbers so that any losses are small and any extra worker costs are kept as low as possible.

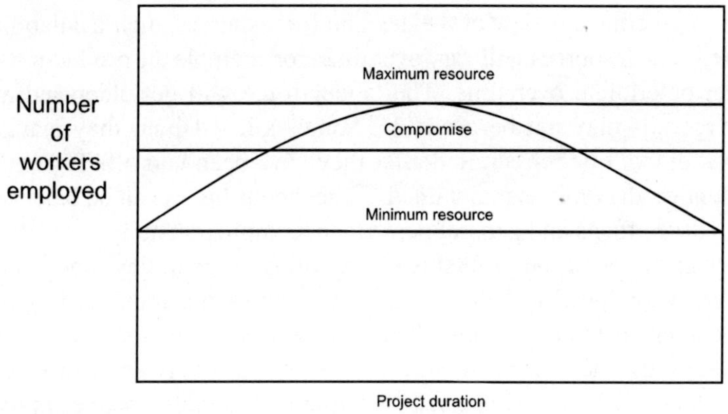

Fig. 16.1 Resource capacity

In the longer term if demand rises, then the organisation is under-resourced for the duration of the additional demand. If demand falls, then there is under-utilisation of the capacity which is available. The cost of either is significant because organisations can not carry surplus capacity which is costing money but not making income. When there are extra resources needed, the use of sub-contracting can also be expensive. As an alternative the workers can have additional training to be made more multi-skilled which allows some labour flexibility. A large organisation may be able to transfer labour between departments or sites.

There are benefits in using sub-contractors rather than having unused, but paid for, man-hours. The potential benefits of using a sub-contractor are clear where there is a lot of uncertainty about the man-hours that will be required. However, too much reliance on this approach can be dangerous Overtime working is another option in the short term, but the ability to do this may be limited by workforce attitudes and the costs of a wage premium. So overtime will potentially be as costly as using a sub-contractor. Of course the additional cost has to be considered in the context of the potential costs of lower production and late delivery of a ship.

The above emphasises the need to have a full understanding of the skills available, the need to expand these and knowing how to do so at short notice.

In the medium term, more pre-planning can be used to sub-contract elements of the work on a contract, subject to the potential cost implications as outlined above. There is also more time to carry out re-training of some of the labour force so that flexibility is available.

In the long term it should be possible to re-structure the organisation, as part of the shipyard strategy development, so that the balance of labour between different activities and skills is changed to suit, for example, a changing product range. It is also possible in the longer term to adopt more automation, replacing workers with machines, although the machines do have a fixed cost. The basis for this and other options is a thorough review of the labour force as described above.

After completing a review of the demand for resources, their availability must be considered. The resources will vary over time, for example the production equipment will often deteriorate over time. The labour force will get older and apprentices or other recruits may not be available. Some skilled labour may leave for better conditions or can't be re-recruited after they have been laid off. To some extent, if a consistent product mix is maintained, some labour losses can be balanced through learning curve effects and general performance improvement.

Earlier in the book the problems of recruiting shipbuilding labour in advanced nations was mentioned and this is in fact an increasing problem in many places. When a country is industrialising, the classic movement is first into heavy industry, then lighter industries such as consumer goods and finally services or electronics. Although that is a grossly simplified description, it broadly operates in most places. Generally people prefer to choose comfortable working conditions.

The result for a developing nation is an initial move from agriculture to heavy industry, giving workers a more stable and usually increased income. The heavy industry is also initially able to absorb a lot of relatively low skilled labour. The fact that the work is labour intensive, often heavy manual labour and is in many cases in the open air is not of great concern to a population used to agriculture. However the start up industry is not very efficient, and as labour costs rise there is a need for greater skills so the industry can become or remain competitive.

If the industrialisation is successful, then lighter industries emerge. These are located in factories and the working conditions are better than in a shipyard, where some of the work is conducted in the open air, in confined spaces and at height. The factory work is also less physical, so more attractive to workers. There will also be increasing numbers of jobs in transport and services, again potentially more attractive. Consumer goods are likely to follow, with factories set up to cater for increasing local demand. Also, large, often multi national companies will be drawn to set up new factories, which can take advantage of a labour force increasingly used to industry and also still accepting relatively low wages. At this point, the shipbuilding industry is likely to find increasing difficulty in recruiting, and importantly retaining skilled workers.

Finally, in a well developed economy increasing demand is found for electronics, luxury goods and services. Often there is still a lower comparative labour cost to attract international companies. Also increasing affluence goes alongside improved standards of education leading to a demand for higher skilled and higher paid work.

As a result of the changes described, the shipbuilding industry in any country may have a limited lifespan and, arguably, as countries change more rapidly and move to newer industries more quickly, the lifespan is shortening.

Other social issues affecting ship construction include health and safety legislation and trade unions. Building ships was always a hazardous industry. In the nineteenth century and into the twentieth century safety was almost non-existent, and several deaths during the construction of a single ship were regarded almost as an acceptable occupational hazard. Legislation to promote safety was largely introduced during the latter half of the twentieth century. Occasional deaths in advanced shipbuilders still occurred, but always became a focus for an enquiry, improved safety procedures and often new legislation.

The effect has been to increase costs, and this is can be a key factor in the industry moving to newer countries. Visiting shipyards in the late twentieth century, there was a strong contrast in many areas of the business. Scaffolding to provide access to high working areas is a good example. In the well established shipyards, in Europe and Japan as examples, specialised metal scaffolding frameworks were used, usually with safety fastenings. Metal bound wood planks were used for the working platforms, with toe boards to prevent slipping off the platforms. Hand rails were a feature, with purpose constructed staircases and sometime elevators. In any particularly hazardous areas, safety harnesses were attached to the workers to prevent falls.

In contrast, in new shipyards in developing countries, scaffolding was often made of bamboo or other wooden poles, lashed together with rope. Flimsy work platforms with no toe boards were provided. Instead of fixed ladders or elevators for very large ships, workers simply climbed the scaffolding to reach their work areas. This resulted in a much lower cost shipbuilding operation, but at the expense of worker safety.

However, the faster pace of change and international pressures are spreading the need for safety round the world more rapidly.

Occupational health is also an issue which has become prominent in recent years. The dangers of hazardous materials, noisy equipment, vibration and noxious gases have been identified and action taken to avoid them. Many processes in common use up to the late twentieth century have been phased out and replaced by quieter alternatives.

When ships were still riveted, hammers and later pneumatic hammers were used, in large numbers, producing a noise level more than enough to cause deafness in later life for workers using the equipment and also those in the same spaces. In the mid 1960s and early 70s, pneumatic equipment was still used for caulking welded seams, producing the same effects. A single piece of equipment in a confined space could make conversation impossible within twenty metres of the noise source.

The introduction of welding to the industry on a large scale resulted in a poisonous atmosphere where the process was in use. Again, confined spaces produced the worst effects, and the operators and those near them were exposed to toxic fumes. Over time, masks and sometimes breathing apparatus were introduced to minimise the hazards. Eye damage was another problem. Although welders were supplied with darkened eye shields, built in to a helmet, others in the area could have problems with looking directly at a welding arc. Improvements in welding and use of shielded arcs have improved the situation. The elimination of some facilities, for example blacksmiths workshops with open coke fires and red hot metal has also improved part of the working environment in shipyards. Vibrations from pneumatic tools could also cause muscle and nerve damage to hands and wrists after prolonged use and where still in use these are much improved.

Personal protective equipment is mandatory in most shipyards. This includes hard hats, safety glasses, ear defenders because a shipyard is still a relatively noisy environment, gloves, steel soled and capped work boots and company supplied overalls.

Small, developing shipyards often had workers in regular shoes, even sandals, with no other protective equipment supplied.

Although the modern shipyard is a far better working environment than it was historically, it is still not a place that many people wish to work in.

For an existing shipyard or for a new development which has recently recruited there is a further question to ask about payment systems. Basically two alternatives are available, first payment for hours worked and secondly payment by results. In the first case, there are set working hours and an hourly payment rate, so workers are paid for attendance. There may be fines for late arrival and typically the wages for half an hour are deducted. In a well managed shipyard, this is a good system as the work will be planned and materials available so the workers stay busy. If the organisation is not so good, then workers may be waiting for some requirements and making no contribution to production. This creates a serious problem for the management and would potentially lead to considering the alternative of payment by results.

This option has been available to managements in all sorts of industries, and agriculture, for as long as these have existed. Basically the workers are paid for what they can produce. So for a shipyard, payment could be for the number of items made, the metres of welding completed or any other readily measurable output. The attraction is that this payment approach should benefit both parties, in that management gains more output and workers have higher pay.

This approach has grown recently with the increasing international competition forcing managements to seek cost reductions. Along with outsourcing, trying to obtain more production for less cost is a driver for many shipbuilders. The result is not always so beneficial. Sometimes the workers will be content with a level of pay and will restrict output, and at other times if the management cannot supply enough work there will be labour trouble. Management may want to reduce the high pay and therefore set minimum output targets which the workers regard as too difficult. In pursuit of high pay the workforce may produce lower quality output than is required. Alternatively, the management may wish to restrict payments by rejecting some of the output not always fairly.

When it works effectively payment by results can be beneficial, but often it is a source of friction between management and workforce. For example the workers do not generally have any control over the supply of materials and interim products. If there is a shortage of these, then no work can be done and the management will not pay. The workers may decide informally to limit their output, so that the wages are adequate, but will not choose to produce more for a higher payment. This is most likely towards the completion of a shipbuilding contract. The workers hope is that the work will be available for a longer period, if they slow down the pace. They may consider it possible that the management will offer a higher rate to ensure completion of a ship on time.

There is increasing international concern about environmental pollution. Concern about the health and safety of workers in shipyards is mirrored by concern for the general public, especially those living or working close to industry. As a result, there is increasing legislation to control or ban some industrial activities which cause pollution. A number of shipyard processes can cause problems. These include surface cleaning, shot or grit blasting of steel, paint coating of steel, waste disposal in general, welding especially where fumes are released and noise.

Environmental legislation is found in most countries, although the levels of enforcement may vary considerably. In most, in order to carry out an industrial activity which generates wastes, it is necessary to obtain a licence from the appropriate regulator or inspectorate. Most environmental management agencies will use a mixture of enforcement to challenge polluting behaviour and assistance to companies to avoid pollution. Some can advise on how to improve operations to avoid any adverse impacts on the environment. Sensibly preventing an incident is far better than handing out a fine.

Assistance in interpreting and conforming to legislation is also usually available for the same reasons. A similar approach is generally used to that for quality management. The relevant standard is ISO 14000 and this is similar to ISO 9000 for quality management in that the focus is on the production processes which will generate no or minimum waste. As with ISO 9000, certification is carried out by accredited third-party organizations rather than by ISO directly. A similar approach is used to review the processes, with wastes rather than products as the outputs.

Establishing ISO 14000 in a company is basically a project. There is first a requirement to review the company and how it deals with wastes and other environmental problems. Then it is necessary to implement the standards, generally with a third party, accredited advisor, as with quality assurance. Once the system is in place, to monitor how it works and the results and if required make adjustments and improvements.

In general the international objective is to ensure that pollution is minimised, or avoided completely. Although the emphasis is on the management of waste or other pollutants, not necessarily on the processes to be used, many of the industrial processes may not be used if there is not an environmental management plan in place. The agency usually has the power to order work to be stopped if the process is not complying with requirements.

Often, companies are required to monitor their own performance, and are responsible for compliance with the legislation. There are inspections from time to time to ensure compliance in most cases. The inspections are often as much about record keeping, rather than checking any actual pollution, which might be considered to a system to allocate blame in the case of unexpected problems, rather than avoiding the problems in the first place. As a result the information on waste management is in reality often more used for investigations after the event than for active avoidance of pollution. However for shipbuilders there is no alternative to compliance.

A problem for ship construction, which it shares with other large scale industries, is that the site is large, the level of activity is high and any problems are readily observed. It is easier to check on a shipyard than on perhaps one hundred small businesses where any non-compliance with legislation is easier to hide. These produce smaller individual quantities of waste and so it is easier to conceal any pollution.

The companies should use appropriate technology in their operations, and must still comply with regulations. Overall the regulations and their enforcement can be expected to become more restrictive. Often the legislation is not prescriptive, that is to say companies may operate in any manner, using equipment which is not disallowed for health or safety reasons, provided they do not pollute. Self-monitoring is the expected practice simply because the regulators are not able to monitor industry on a continuous basis.

Noise is one form of pollution which can affect local communities and enforcement is often carried out local authorities. Noise pollution is primarily a health and safety issue, but also affects the local environment. It is a particularly important problem at night when people are trying to sleep and the level of ambient noise is low because there is for example no traffic on roads. It is the most common cause of complaints especially from local populations about any industry. Depending on local and national regulations, a large number of complaints can be problematic for a shipyard management.

Noise sources include services, with generators, compressors, pumps and frequency converters as a large part of the problem. Services are also a source of leakages, and noise can be generated at end use consumption.

Noise can be controlled by choice of quieter processes, probably the major example in the past being the elimination of riveting which was responsible for large scale deafness among shipyard workers. More recently, the elimination of pneumatic hammers has also reduced noise levels. Management of the noisy processes in use can reduce the effects locally, by sound proofing and as a last resort restricting operations to daytime when the problem is less noticeable.

Surface cleaning and blasting is necessary to prepare steel surfaces for over coating. A range of processes is available, including several which are no longer in general use for ship construction, though may be useful for dealing with small areas of steel. Mechanical processes, either portable machines, sanding, grinding or even small hand tools fall into this category. Grit blasting contained in a small, moving cabinet is also limited to small areas of repair.

Blasting is not carried out in the open air, or if absolutely necessary can use water injection to prevent airborne pollution. High pressure water blasting for washing and ultra-high pressure water blasting for surface preparation avoid airborne pollution but are also normally used mainly for ship repair. Dry blasting with grit which is then recycled is the preferred method for steel cleaning. At the start of the ship construction process the blasting equipment is contained in a cabinet, through which the steel plates or profiles can be moved on a conveyor. This prevents any pollution and also allows the grit or shot to be recovered, filtered and much of it re-used once or twice. Spent material is collected for disposal.

For finished units and blocks, the final cleaning is ideally carried out in a large building which again contains the process, and the blasting material is reused and then recovered in the same way.

The preferred application option for coatings is airless spray. It is fast, provides a good quality of painting and is safe. Ideally the work is completed except for unit joining areas in a special building, not left to the open air or the building dock. Outside the problems are associated with overspray, where windy conditions increase overspray, and paint particles may carry long distances. Operator training is very important to manage the quality and to minimise the wastage.

Welding is primarily a health and safety issue. Avoiding fumes close to the workplace is important, by means of extraction systems. These can be within workshops, in work locations on the ship, or integral with the welding equipment. Air-fed helmets are used for confined spaces and over time it may be more possible to replace people with more automation, or to use remote control where possible.

Waste disposal is important, both to maintain a clean and tidy work space and avoid hazards and also to manage the large quantities generated by ship construction. Reasons to reduce waste include avoiding the high disposal costs, in landfill or by incineration. There are increasingly strict hazardous waste regulations and generators of hazardous wastes are liable for their waste even after it is transported somewhere else for disposal. Waste reduction is perceived favourably by the community and can enhance public image. Waste is also a risk factor, if anything does go wrong.

A company should encourage employees to assist in identifying wastes. This includes leftover materials, use of energy and outputs including products and waste. A review of what is being disposed of or released into the environment is required. There are options to reduce waste, to reduce the sources by changing materials where feasible, choosing materials which can be managed on site, improve the processes in use and improve the housekeeping on site.

They should also make arrangements to contain any spillages, make use of off cuts of pipes and steel, inspect the site for leaks in services and remove any waste left around. It is also possible to train employees in better materials handling to avoid breakages, and to maintain materials in protective environments. Immediate collection of any waste can avoid a hazardous waste contaminating large quantities of other materials which might be useable or saleable. Mixed wastes are more difficult to treat and/or dispose of, and may make otherwise reusable waste useless.

It is important to reuse any waste where this is possible and otherwise to recycle waste. Often a specialist sub-contractor is used to dispose of the wastes. The workers are encouraged to sort waste at source into containers which can then be regularly collected. Different materials are kept separate so they can first be reused if possible. An example is shotblasting materials, which are expensive. The used material is filtered to remove debris and then sorted by size so that a mix of new material and clean partly used material is used. This gives very effective cleaning. The material and debris which are not reusable are then sent for disposal.

Dealing with the shotblasting material is largely automatic, as part of the steel preparation process. Using the same objectives for other materials at all stages of production for all activities is more difficult. However if work study is applied, the value of careful sorting and reuse or disposal can be decided and there is potential benefit to the shipyard.

Chapter 17
Progress Monitoring and Control

If everything seems under control, you're not going fast enough
—Mario Andretti
You may delay, but time will not
—Benjamin Franklin

Once a shipbuilding project is started, it is essential that processes are in place for the monitoring of progress. Even if progress is exactly according to plan, which is often not what is actually happening, it is necessary to know that for certain. More importantly if progress deviates from the intention, and that is usually the project slipping behind the schedule which was planned, that must be identified.

So it is clear that monitoring progress is an important element of project management. The timescale for a ship or offshore project is generally measured in months and years. As a result, if any delays occur and are not corrected, which includes underlying problems causing the delays, this can result in the project going late. It is therefore essential that progress is monitored, and in such a way that any possible deviation from the plan is detected early enough for corrective action to be taken.

The project will have been planned with several levels of detail. For the purposes of this book, four levels have been identified, but more or less may be found depending on the organisation doing the planning and the complexity of the project undertaken. The four stages or levels of planning are Corporate, Strategic, Tactical and Detailed, and have been described elsewhere. Monitoring is carried out at each level.

At the corporate level, monitoring is based on major events in a programme of ship construction which may have many projects in it. Typically the period covered is five years, though this can vary, again depending on the number, size and complexity of the projects in the company.

Monitoring is primarily for the senior management of a shipbuilding company, who are most concerned with financial achievement. Because the structure of shipbuilding contracts generally provides for progress payments on the completion of significant stages of the ship design and production, these are the basis for progress monitoring. A very clear point in a ship project is the delivery date, which usually provides a large payment. Other key events in the plan are for example, completion of design work, purchase and delivery of steel, a first unit into the building dock and float out from the dock.

G. Bruce, *Shipbuilding Management*,
https://doi.org/10.1007/978-981-15-8975-1_17

The strategic plan for a shipbuilding project is typically a one hundred event network. The number of events can vary considerably, but for a medium sized cargo ship, such as a handy-size bulk carrier, one hundred is a reasonable number. Using the network, the progress of the project is monitored by recording the completion of planning units (key events) in accordance with the schedule. It is important also to record the expenditure of man-hours which should be in accordance with resource loading.

At this planning level the focus is on the individual project rather than the whole shipbuilding programme and the progress monitoring is primarily for the project manager to be able to check on progress between payment milestones.

The tactical plan is the three month plan for an activity within a function, usually a department of the shipyard and is to check the progress of the interim products that the department has scheduled for the period. The three month plan is a rolling programme which is updated at least monthly with new interim products. These products can be for any of the current projects in the shipyard, so for a large operation can include several ships. The progress is monitored by successful completion of interim products in the correct sequence and on time. At the same time the departmental man-hour expenditure is measured.

Detailed planning is based on work packages in individual work stations. The timescale for the detailed workstation schedule is typically up to one month and this is updated usually weekly The progress is monitored largely by recording the completion of each work package within scheduled time constraints.

It is important that the progress monitoring is focussed on critical tasks and major deliverables. Critical tasks are generally those on the critical path of the project network, which are those tasks where failure to deliver will affect the entire project. It is not necessary, indeed it is usually undesirable, for the senior and even the project management to know about the status of all the tasks for the purpose of progress monitoring. What matters is that they are informed if anything is late and may affect the whole project.

However, the management must still be able to review overall progress, and if this is unsatisfactory be able to look into the detail alongside the local management in order to identify causes and be able to find solutions. So the total work done and expenditure of man-hours is recorded.

The need to plan the project on the basis of deliverables has been emphasised and it is equally important to monitor on the same basis. The project plan, based on the hierarchical work breakdown structure, has deliverables at each level of planning. The levels of planning, and progress monitoring, are linked. Some examples are given here:

At the corporate level, a major event for a ship is keel laying or first unit into the building dock. This is a deliverable of some importance, since it marks the effective start of ship construction: It is also usually the trigger for a stage payment on the contract. The management of the company will record that the due date is met and therefore the company is paid. It has been known for shipyards to place an unfinished steel unit in a dock in order to claim a payment.

For the project manager, at the strategic level of planning, the objective is to achieve that keel laying date. The key to this is the timely completion of the first hull structural unit or block. At the end of the sub-network for the unit is its completion date. The unit appears on the main project network and satisfactory progress has been made if the unit is ready on time. This will usually be some time before the date it is due in the dock.

At the tactical level, the completion of all sub-assemblies for the unit is important. If there is a delay in any of the sub-assemblies, then that provides an early warning to the management of the production department that the unit may not be ready on time. This alerts the project manager to a potential cash flow problem. So the measure of progress would be that all the assemblies and other elements of the unit are ready on schedule.

At the detailed level, to take an early stage of part preparation, the cutting of all components for particular work packages is required. So progress is measured by the completion of a cutting work package on the date required. If there is a delay, then the information will be passed to the local manager and if necessary to the ship project manager.

The success or failure in meeting dates at a lower level of planning is therefore an early warning of probable success or failure in meeting higher level dates. So the delay in cutting some steel plates will warn of a potential delay in assembling a unit. This in turn warns of a possible delay to keel laying and hence to a progress payment. The point is that the initial warning is weeks or even months before the payment issue, and this should give adequate time to make corrections and catch up. During the process of determining why a work package is late, any underlying problems should also be uncovered, which leads to action to correct these.

Progress monitoring compares the reality of production with the plan, and this is done by measuring not only work package completions, but also completion for example of information availability, material availability and testing. Expenditure is also monitored measuring the man-hours spent and any other costs associated with production.

It is necessary to measure the work actually completed in order to monitor work progress. It is common to use percentage completion as a measure. That is, the project management takes estimates of percentage completion from the various parts of the project and uses these to assess overall progress. This has some dangers.

Basically most managers and all supervisors tend to be optimists when progress is to be considered. This is in contrast to when work is being planned, when they are pessimists and ask for more time to be made available. As a result, percentage progress estimates tend to be over-optimistic initially. Progress is recorded steadily, but then the work completed is often recorded as "90% complete" for long periods. On occasions a need for rework may even result in completion percentage being reduced.

If the timetables set for tasks are long, then progress on them may be reported as "well on" or "90% complete" at several progress meetings. This may be satisfactory, but it is more likely that it may conceal serious problems. The problem for the project management is that there will be uncertainty, and either the progress as reported is accepted or if not then the management will have to investigate. This can be a serious burden, and distract them from more important matters. It is also important to focus on tasks which are on the critical path, since these will determine actual overall progress with the project.

There is another danger at the detailed level, especially in component production. The measurement of work done in, for example, plate cutting or pipe production can be by volume of production as well as by completion of specific work packages. There have been cases where senior management showed more interest in the numbers of components produced than in which components were produced. As a result, a lack of overall progress was masked until shortages appeared in assembly. That is, there were large numbers of items being produced, but many of those could not be used in assembly because they were produced too early. Others were produced late, delaying the whole project.

The scheduling of work in component production may also be influenced by a desire of local management to see all the people and equipment "busy". Where set up times are long, as in the case of pipe production, work is often performed out of sequence in order to maintain production rates. This is acceptable as long as due dates are also maintained, although this can cause problems with the generation of production data from the designers. It also usually results in large numbers of pipes in storage because they have been produced early.

Bearing in mind the pitfalls of monitoring progress which have been identified, it is safest to monitor work packages at the detailed level. These work packages should have short timescales, ideally no more than one week in duration. The timescale should also ideally be less than the interval between progress reports, and then all work packages are at either of two stages of progress. They are not finished, in which case the progress is recorded as 0%, or they are complete so progress is genuinely 100%.

It is perhaps reasonable to record some work packages as in progress, although this can be started but still zero per cent complete. Or they can be assigned a completion of 50%, but it is usually better to split potentially long duration work packages into two or more parts so the recording can be maintained as 100% or 0%.

Typical work packages ideally have enough work for up to five workers, depending on the activity, for one week. This will equate to about 200 man-hours. The duration is the most important aspect, and keeping this short allows potential problems to be picked up quickly. This very simple approach to recording progress avoids over-recording progress and focuses attention on work that has very recently fallen behind schedule. The duration is the most important aspect, and keeping this short allows potential problems to be picked up quickly.

Aggregating work package progress gives tactical and strategic completion status, because the total progress is still important for senior management. So the number of work packages at 100% completion is summed and this is overall progress. It avoids a situation where in an extreme case it might be possible to have 90% of packages at 90% completion. This would indicate 81% progress overall, but because the packages are not complete then the next stages of work for which they are required cannot progress.

Provided the work packages are small, the recording of progress is realistic, and will give very similar percentage overall progress to using work package percentages. For a medium sized ship, the numbers below are typical, but the number of work packages will reflect the size, complexity and timescale of the project. Around 20 work packages go into a planning unit (event on the project network) and around 100 planning units make up the ship, so the project is being monitored in terms of some 2000 work packages.

The second part of progress monitoring is to monitor expenditure of man-hours, where the total worker hours for a project are the sum of hours for all the work packages. Ideally, the hours would be spent at a constant level. This can be achieved in a factory using flow lines with a fixed work force but is not generally practical for a large, made-to-order product such as a ship. The major shipbuilding exception is in the part manufacturing and some sub-assembly areas, where the man-hours should be fixed by the staffing of the work station.

The worker hour expenditure for a project generally follows a pattern with a build up from zero to a peak level at the start of the project, then a steady peak level for most of the time followed by a run down to zero towards the end of the project. The estimated hours can be presented in this form using information from past projects as the basis for estimating the resources needed for a new contract. For a similar ship to past products, a curve showing the variation in numbers of workers required over the life of the new contract can be created. Then by taking the estimated total hours and the overall timescale and applying these to the normalized historic curve the numbers expected to be employed at any time during the contract lifespan can be found. This also offers a means of checking that the correct staffing level is in place for a contract at any time.

The man-hour expenditure over the duration of the project can also be shown cumulatively. The cumulative expenditure is in the form of an "S" curve which is illustrated in Fig. 17.1. This is a very common presentation of the same data. The S curve shows the total hours to date at any time during the project. It can be represented as the percentage of total planned hours, against percentage time expended. The actual man-hours can be plotted against the plan.

The question is then how to interpret the information. Monitoring of progress informs the project management whether the progress to date against the expenditure to date is in accordance with the original plans. If there is a deviation from the plans, then there must be an investigation to determine what is really happening to the project.

Spending fewer man-hours than planned can have several causes. The first alternative is that there are not enough workers on a project, which can indicate an overall shortage of resources, often because those workers required are still working on an earlier project which is behind schedule. Management is usually keen to see completion of a ship because of cash flow requirements and often move workers to achieve a completion.

Another possibility is that there have been late deliveries of materials so work cannot proceed, and again workers are re-deployed. There may also have been production problems, causing delay.

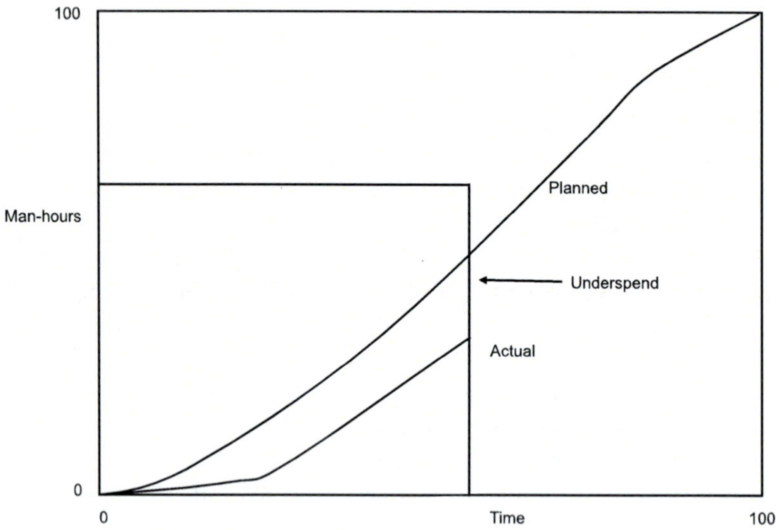

Fig. 17.1 Man-hour expenditure

A further possibility is that the performance has been better than expected and planned, work has been accomplished using fewer man-hours than was planned and potentially a large profit can be made. This may be considered to be the least likely reason.

The effect on the project, and the possible outcomes, can vary. In the best case, the project will finish on time, and at less cost than planned. More likely is that the project will be late, because there are too few workers. Depending on the extent of the deviation from plan, the project may finish very late, because of serious production problems.

In order to be able to predict the likely outcome of a project, both work progress and man-hours must be considered. Then the expected progress against planned man-hours can be plotted. The example in the figure indicates man-hours more than planned. It is also important to monitor progress against time, because then both actual man-hours and actual progress can be plotted.

Once the project management has access to reliable progress information, then the method to calculate and display progress is usually known as earned value management. This is a project control method which measures project progress on the basis of the costs that have been incurred, the time elapsed and the work completed, all of this compared against a defined scope of work.

The project manager needs to know several pieces of information in order to maintain control. First the manager needs to know how the project is progressing, what has been achieved for expenditure to date and what the final cost is expected to be. He should also want to know whether the progress reports he has received are reliable, how efficient is the project and is the customer happy.

Figure 17.2 shows a more complete presentation of the status of a project in terms of time elapsed, work completed and expenditure. This is a basis for earned value

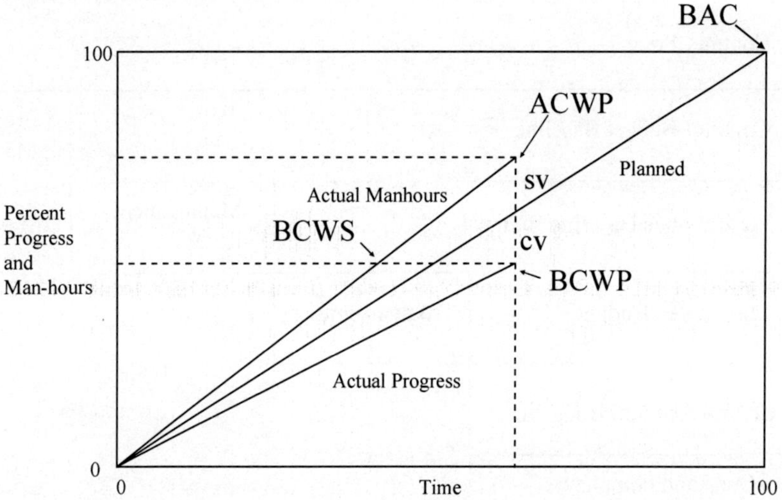

Fig. 17.2 Earned value management

management. This starts with a definition of the project scope of work, which is the baseline for cost, schedule and technical performance. There must be a number of features for the progress monitoring to be effective. These are objective performance measurement reporting, variance reporting and correction of deviations. Sometimes making timely changes to the baseline is considered important, which is a potential problem because changes to the baseline might imply a change of plan which is problematic in ship construction projects. Ideally the plan is maintained to achieve the final delivery date, unless revisions which are contractually acceptable have to be made.

The three key data elements are the budgeted cost of work scheduled, BCWS, which is what the project expected to achieve. Then second the actual cost of work performed, ACWP, in other words what actually happened. Finally the budget cost of work performed, BCWP, which states what was actually achieved.

To start with the plan, the budget at completion, BAC, is the budgeted cost of work on completion. That is the expected cost outcome of the project if all the work proceeds to plan and to budget. The timescale shows when the project should be completed.

The comparison of progress achieved and expenditure is made by using two earned value factors which are cost variance (CV) and schedule variance (SV).

CV is equal to BCWP minus ACWP, where a positive value indicates that the project is below the planned cost. The cost variance can also be expressed as a percentage. For clear presentation of the status of a project, the most useful measure is the cost performance index (CPI). This is simply calculated by dividing BCWP by ACWP. If the project is exactly on track, then the value of the index is 1.00. If CPI is greater than 1.00 then the project is running efficiently and under budget. However, if CPI is less than 1.00 then the project is running over budget.

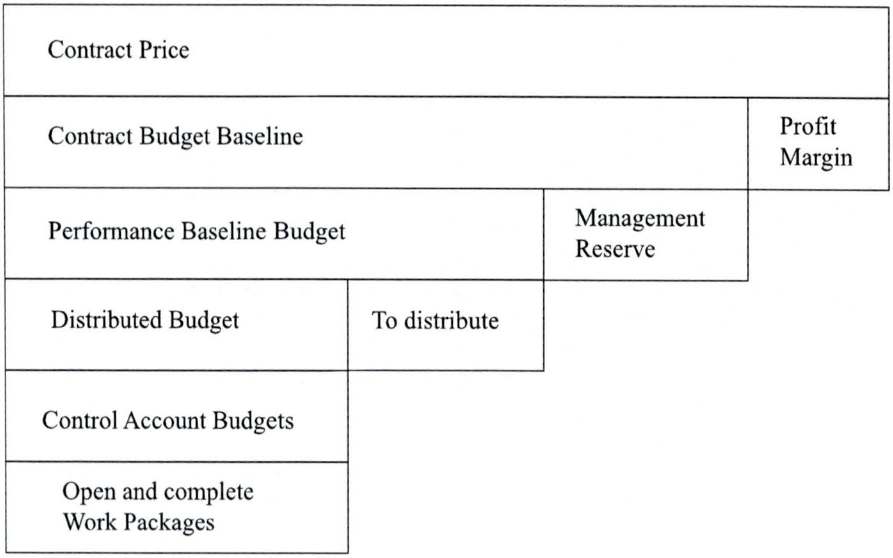

Fig. 17.3 Budgeting

The schedule variance SV is equal to BCWP minus BCWS where a positive value indicates that the project is ahead of schedule. The schedule variance can also be expressed as a percentage, but again the most useful presentation is to calculate a schedule performance index (SPI). This can be calculated by dividing the BCWP by the BCWS. If the index is greater than 1.00 then project is running ahead of schedule and if it is less than 1.00 then project is running late.

In both cases the indices can be plotted so as to highlight the deviation, if any, from the planned outcome. This is a clearer presentation than using the actual values.

To use earned value, the initial requirement is to have a robust and credible project budget, which is divided into measureable packages of work. As stressed earlier these should be of short duration and similar in size. The next requirement is a reliable method of measuring the quantity of work which has been completed as discussed earlier in this chapter. Also required is a sound method of collecting costs incurred for each measurable package of work.

From the above it is clear that the work breakdown of the project is critical to its success. There is an initial breakdown into main elements, then a further breakdown into sub-elements. These sub-elements, broadly corresponding to the interim products for a shipyard, allow for the allocation of tasks to specific work stations or work groups. This is the basis of project planning. Budgets can then be set, generally in terms of man-hours to be spent on a task and any other resources used. The budgets are based on the allocation of the estimates made of the project cost. Cost is allocated for each interim product and there is a hierarchy of budgets which corresponds to the work breakdown structure for the project.

The budgeting process, shown in Fig. 17.3, is carried out in parallel to the work breakdown structure and planning. The makeup of the budget has a number of parts, which are designed to make the project manageable. This starts with the price to the customer, which is what has been agreed should be paid to the contractor. The contractor has a budget baseline, which is what the project should cost, ideally with a difference between that and the price which will provide a profit.

There is then a performance baseline, lower than the budget baseline, which is what the contractor hopes the project will cost. The difference between the budget baseline and the performance baseline is often termed the management reserve. This is a budget element which can be made available if the management agrees that the performance hoped for, and hopefully planned for, cannot be achieved for a specific part of the project.

Setting the performance budget baseline requires very careful thought. Too often the management of a company will try to set an arbitrary reduction percentage across all the work packages in a project. Unless there has been careful consideration of how the reduction in cost can be achieved, this is more a pious hope than a realistic expectation. There is also a serious danger that the whole planning and budgeting process will not be credible to the lower management and the workers. The production function in a company will always ask for a higher budget than is really needed, so they can have a contingency to manage any problems in meeting the plan. If they are asked to perform better than in the past, then they have to know how an improvement can be achieved.

The total planned work packages constitute the work which will result in cost for the project. These planned packages each have a budget, and those where work has been authorised and cost committed are the work in progress. For these authorised packages, the cost is allocated to a specific account for cost recording.

Where work has not yet been authorized, there is undistributed budget which has not yet been allocated to a cost centre. Nothing can be charged to these work packages until they are authorized. Then once a work package has been completed it will be closed and no further expenditure is permitted against the budget cost. This gives the management control over the expenditure, and if the work packages are of short duration, only a small number are open at any one time making the control easier.

The production work packages, which are the majority, are discrete and specific tasks in the project, and must be measured by their state of completion. The usual descriptions of earned value allow for percentage completion of work packages, which this book regards as a potential problem. However if the work package is only recorded as not started, started or completed and also the resources which have been expended are measured accurately the method is sound.

There are some potential problems in recording the cost and progress of a work package, again especially if the package can have a percentage completion recorded. If the work package is over budget then it is likely that the supervisor responsible will request more resources. On the other hand if the work package is below cost the supervisor is more likely to keep quiet and try to keep the unused resource as a personal contingency. There is also always potential for cost misallocation where packages are open for any length of time.

It is common for a customer to request additions to the work scope, in the form of variations. These must be considered to be extra work, based on the terms of the contract, and a price for them negotiated. It is not appropriate to use the management reserve for this purpose. If there is work unforeseen by the contractor, which must be done under the contract terms, then the management reserve can be allocated.

The potential effect of a variation on the project must be evaluated. There are many reasons for variations, apart from a customer request, which will result in some work packages taking more man-hours than was initially estimated and planned. In an extreme case, some work items may have been missed entirely from the project plan. The need for evaluation of any potential extra work is because any variation can affect the overall project cost or timescale. If a variation is acceptable to the contractor because an extra payment is available, and acceptable to both customer and contractor because there is no impact on timescale, then additional work packages may be agreed. Alternatively, extra resources may be allocated to existing work packages, and these may also be given a longer timescale for their completion.

It is necessary to review the possible effect of additional work. First it is necessary to consider whether the new or revised work packages are on the critical path of the project. If they are not, then it is possible that the critical path may change. If the project timescale is likely to be affected, then the change may not be acceptable. Possibly additional resources can be used or in an extreme case the project plan may need to be revised.

All variations are potential risks to the project. Any variation request must be recorded and the potential risk must be reviewed to determine the potential risk to the schedule, the cost associated with the variation and above all, who will pay.

Progress meetings are important in a shipbuilding project. Meetings have a peculiar status in business, and almost everyone claims to dislike meetings, claims they take up too much time and that they don't achieve useful results. On the other hand, people do attend them, because in fact meetings are essential. They attend to find out information, agree things to do and to avoid being assigned unwanted tasks in their absence. Meetings do fulfill an important purpose, but they are very often managed badly.

However, if meetings are set up properly and managed effectively, they can achieve a lot of useful work in a surprisingly short time. The ideas here have proved to be useful in improving meetings, by reducing their duration and increasing their value.

Formal and informal meetings are a major feature in almost all organisations. In general, a meeting is held for one of the following purposes, first to receive information, usually from someone senior to a large group of employees. Smaller meetings are required to exchange information between participants, for example progress on a project with various partner organisations. Meetings are also needed to review information and reach some conclusion, such as a decision on whether a project should proceed in the first place. And meetings can be used to decide on a future course of action, for example a board of management choosing a future corporate strategy.

Meetings are an inevitable part of business life, and they are very important for complex projects because they are the means of communication between different organisations, companies, departments and individuals. The key to a successful meeting is how well it is organized, in preparing for it, in its conduct and in carrying out agreed actions. There are a number of aspects of formal meetings which, when used properly, can make it successful. Even an informal meeting is best conducted within a basic structure.

As a minimum, the preparation for a meeting should include a note for each participant, made individually in discussion or circulated, giving advanced notice of the meeting. This informs each person of the time, the expected duration, the precise place to meet, who is expected to attend and an agenda of items to be discussed. If any complex item is to be discussed, or any detailed notes will be reviewed at the meeting, then it is essential to also circulate these to meeting attendees in advance. This saves time because everyone has had an opportunity to study the issue and is prepared for discussion.

If the item is to be presented at the meeting for information, then careful consideration should be given to the form of presentation, and a note of the key points provided for participants to take away from the meeting.

For discussion and decision making meetings, their conduct must be controlled to make sure the agenda is completed, and to make sure any agreed actions are assigned to be carried out. The minimum requirements are a designated chairperson to control the meeting and a secretary to take minutes, in particular to record decisions and actions. During the meeting, a few ground rules help to speed things up and keep to the point.

Everyone should arrive on time, because lateness is a frequent problem with meetings, and stay to the end. When the meeting starts, first record the attendees, agree the agenda is correct, and see if any additional items should be discussed. Revise the agenda if necessary, for example if an urgent problem has arisen or if key people cannot attend. Agree the timetable for the meeting, giving more time to more important items.

Everyone should keep to the point, with no diversions. They must participate, not just sit and listen, or worse, not listen, or even worse complain later in private about what was said. Rather than just oppose someone else's idea they should make a constructive counter-proposal. Conclude the meeting on time and if it is necessary, set up a further meeting to resolve any problems which have emerged or not been solved.

The chairperson ensures that everyone has a fair hearing, and ensures that only one person speaks at any one time. The chairperson keeps the discussion within the agreed timetable and then finishes with a summary of the agenda items which were discussed, the agreed actions, and who is to carry them out.

The secretary ensures that everyone has received any required papers before the meeting. The secretary also notes any proposals, summarises discussion, notes down the agreed actions and who is to carry them out. After the meeting, the role is to make sure everyone has a copy of formal minutes as soon as possible, and to be sure at the end of the meeting that everyone knows what actions are required of them.

The above procedure may seem excessive for a small, informal meeting, and it can be simplified. However, the organisation need only take a few minutes, and will ensure that the meeting is productive. It will almost certainly make the meeting shorter than would otherwise be the case. It will also help to ensure that the agreed outcomes are clearly understood, and increase the probability of actions being achieved.

Time is perhaps the most critical resource in many projects so management of time matters. It is always worth reviewing how time is managed by a project team for any project and the project manger should always encourage good time management. Basic questions to consider are how well is time being managed, is it possible to identify where time is being wasted and more importantly, is it possible to identify how to reduce the wastage?

Having enough time for all the tasks to be carried out is a major problem for most people and projects, and ways of reducing wasted time are valuable. In all industries, and in many other organisations, there are usually conflicting demands on the time of managers, supervisors and engineers, As a result, the ability to manage time is an important skill, to help staff to meet deadlines for different activities and tasks.

There is a need both for discipline, to complete tasks within short timescales and not leave them until there are too many for the time available, and for organisation, which is to arrange blocks of time to carry out tasks according to priorities. The simplest technique in time management is to keep a basic time log in the form of a diary, with the day split into half hour periods. Specific timetabled activities, such as meetings, are noted and it is helpful to record actual attendance. Recording what actually took place as opposed to what was planned is essential.

It is also instructive for managers to walk through workshops and other work areas and simply count the numbers of workers, specifically the numbers doing useful work and the numbers who are wasting time in some way. This will give a picture of how an organisation is really using its time through a period of usually one week. The results may be a surprise, but should help to organize time more effectively. More formal work sampling can often reveal opportunities to improve the use of time.

Also it is helpful to assign a priority to each activity, whether it is essential (must be done at a specified time), or important (must be done, but has some time flexibility) or non-essential (really unnecessary, or can be done any time). Over a few weeks, the time utilisation should improve, as management becomes more aware of how their staff are really spending time.

Chapter 18
Management Organisation and Information Systems

The only thing we learn from history is that we learn nothing from history
—Georg Wilhelm Friedrich Hegel
The biggest single problem in communication is the illusion it has taken place
—George Bernard Shaw

When presenting the activity map, what happens in a shipyard was presented as a series of project stages and a number of operational functions. These identify a set of activities which have been the basis for most of the discussion of shipbuilding management. It was emphasised that the activity map represents what happens, not how it might be organised. This chapter will now consider that organisation.

Management is not new, nor therefore is organising the work of a company. There is an illusion that the use of computers has somehow changed everything, and it has primarily in that data can be transmitted quickly. So improvements in efficiency are possible, but are certainly not guaranteed. In the past organisations had to adapt to situations where weeks, months or even years could elapse between a decision and an outcome. Despite all the handicaps, projects were carried out and there is no real evidence that fewer were successful projects than now.

In developing a shipbuilding project, there is a balance between functional and project management. This is indicated on the activity map, with stages of a project and activities within major shipyard functions. The historic functional management is system oriented, and deals with the customer specification, ensuring technical excellence of the project, the overall quality of output and providing technical support to the project. It is focussed on the specification of the ship and on completion on ensuring the completed ship performs according to that specification.

The company can be viewed as a system, with the departments and functions as sub-systems as shown on the activity map. Each sub-system has its own objectives. There may be tensions between sub-systems, because of conflicting objectives. For example the design function may have ideas about the "best" design which do not accord with the needs of production to be efficient. One project manager may consider his particular project to be the most important and seek to obtain resources ahead of others, possibly more profitable to the company.

G. Bruce, *Shipbuilding Management*,
https://doi.org/10.1007/978-981-15-8975-1_18

Such tensions may continue down to a lower level in the company, so that there is a conflict between the wish of a department head to be efficient in terms of total productivity and a project team requiring specific deadlines to be met. So one activity may attempt to optimise its operations, which can be at the expense of others. This is often not helpful, and it may be acceptable to have lower efficiency in one sub-system to benefit the whole organisation. This is known as sub-optimisation, and is not always going to be popular with a department manager. The existence of a hierarchy of objectives can avoid the tensions by clarifying what is the priority for the company as a whole.

How best to manage an organisation has always been a difficult issue. Various forms of organisation have been tried. Some writers have three basic forms, others go up to ten. The basic possibilities seem to be project organizations, process organisations and matrix organisations. Then for larger companies the organisation can also be based on product divisions, market divisions or geographical divisions. Structures within the company have been described as hierarchical, flat, circular or a network.

Returning to the activity map, the first three offer either stages of a project, that is project organisation, or functional, that is according to the processes carried out or matrix as a mixture. If a larger company has divisions, then within those the organisation is still one of the first three for an individual operating company. The three structures offer different ways to picture the company and are really about the internal communications.

In the past shipyards were much simpler organisations, constructing much simpler ships. The fundamental division was into a technical function for ship design and a production function for building. The technical function would include purchasing and commercial activities, usually closely supervised by senior management. The division between the two was also partly based on stages of the ship project, so the technical work would finish and then the production information would be developed within the production function.

There is still no ideal organisation structure for ship production because any alternative will create anomalies and potential problems. However in general it is better to tend toward project oriented rather than discipline oriented, because ships are generally complex products, built in a short timescale, with a fixed delivery date. This describes a project. There is still a need for management of the functions, because most of the workers are not project managers but technical specialists. They require a structure within which they can be trained, developed and guided, as well as having a system for promotion.

So the management organisation has to be some form of matrix, which can be developed from the activity map. Exactly where the boundaries between departments are placed will depend on the size of the company, the complexity of the product and the people who are available.

How the departments are managed is to a large extent an issue of communication. As a principle, the more any individuals or groups have to communicate, the closer they should be in the organisation structure. Looking again to the past, larger companies were hierarchical. Their organisation was based on functional divisions into departments, with several tiers of management. A function such as design would be split into several departments, such as hull, deck outfitting, cargo hold outfitting, machinery space outfitting, accommodation (ship crews were larger and ships often carried passengers) and others for more specialised ships. For the technical specialists, the structure would provide good support and a lengthy promotion ladder.

Communication between these often required large meetings, and problems were referred up the hierarchy to a management level meeting for resolution. The hierarchy is not good for decision making and efficient production. Smaller shipyards, and others more recently, usually have a flatter hierarchy, hence a flat structure. This shortens communication routes, speeds decision making and usually encourages local decision making between different departments. The drawback may be that there are no promotion routes which might lead to higher worker turnover, and sometimes an overload on the fewer senior managers.

Circular, flat and network structures are attempts to show the communications between groups, rather than a radically different structure.

Some successful shipyards have adopted a form of matrix organisation. Generally the pre-contract stages are centralised, as is the completion and delivery of a ship. The bulk of the later design and production work is devolved to local management, based on a project approach. The use of information technology does assist the management of a relatively complex structure, though many of the best management processes predate computers.

For a new ship, the conceptual and preliminary designs are developed centrally, and the preliminary work breakdown structure is also developed centrally as a basis for the build strategy. There is an overall project plan for the ship which is also centrally developed to fit with the overall shipyard work schedule. At an early stage it is important to identify a project manager, who is designated to run the project if the contract is won. The manager should be part of the customer discussions and the project development process to ensure there is clarity about all aspects of the work. Once the contract is signed, before that point in many circumstances, the project team should be assembled so that again there is clarity of objectives.

Once the project is confirmed and work preparation is ready to commence, the organisation of the rest of the project work should be based on integrated, multidisciplinary teams. These are responsible for a specific part of the ship, for example machinery spaces, or cargo handling systems, based largely on the work breakdown for the ship. The design integrity is maintained centrally, and the basic decisions on such aspects as services routes and equipment locations have already been made so the local teams operate within a clear design structure. The teams also operate within a central planning framework, which sets out when interim products are required. A typical team will have a design supervisor for the remaining functional and especially transition design and for all the detailed design. There is also a production supervisor who will manage the relevant assembly and installation.

There is an individual or small group to carry out the integrated detailed planning for the structural units or zones for which the team is responsible. The team will have integrated into it the necessary staff to perform material requisitioning, and these are supported by central materials management who will place orders or supply from stock what is required. This may be internal or external so that control is maintained of the stocks of materials and supplies.

Under the overall supervision of the project manager, the team will order the materials and parts needed. The preparation of materials and parts will be carried out by the production departments, which may have several projects in hand, and the procurement by the materials management function. Once the assembly and installation begins the project teams will have control. If several projects are in progress, then the assembly function should also be under central management. Other exceptions to project control may be services such as quality assurance, materials handling and expert services, for example welding technicians, who need to work across several projects.

The objectives of the organisation are to ensure that detailed design information is produced to meet production requirements. The procurement function obtains items of equipment to a timetable driven by production requirements and also to a timetable dictated by manufacturing. The manufacturing function makes parts to timetables dictated by assembly and installation requirements. Also, units are assembled to meet installation requirements and all testing is completed when needed by production activities, all to meet the project plan requirements.

The various teams have total responsibility for the completion of their elements of the project. The work breakdown is based on the most appropriate for the particular project. For a passenger ship, this is often on the basis of zones (often fire zones) on the ship, because the man effort is in installation of outfitting. For more basic cargo ships, the focus is mainly on the steel structure, except for the machinery space and accommodation block.

In setting up the project organisation structure, the responsibilities for the various deliverables for the project are given to specific management and work teams. This is intended to align the objectives for the teams with those of the overall project, and to ensure correct deliverables, especially when several disciplines are engaged in their production.

In developing the project planning, the time when each deliverable is required to be produced is the basis for the plans and schedules developed for the project. As part of the project management processes, the production deliverables will define the structure and timing of technical information which has to support production requirements. They allow the procurement of materials and bought in equipment, so that purchasing to be carried out efficiently. It will be clear that communicating accurate information in a timely way is fundamental to success.

Management information is therefore an important component of the successful operation of any business, and in one as complex as ship production, it becomes critical. Management can be defined in many ways but a useful definition includes "co-coordinating groups or individuals towards a common goal" and "organising, coordinating and controlling resources to achieve an objective". Both of these statements fit closely with the idea of project management.

Information can be defined as "processed data". Raw data, collected from the plans, schedules and operations records of a business, is not necessarily useful for management. Processing the data puts it in a useful format, by for example identifying anomalies which require some investigation or detecting trends which may be good or bad.

Information can reduce uncertainty, giving management confidence that the actions they are planning or actually taking are the best ones. Information can also be false, or inaccurate, perhaps if it is provided by a third party stakeholder in a project. It can be new or incremental, adding to the knowledge of a management team. It can confirm or sometimes correct past information.

A system can be defined as "a set of interrelated components". The components are interrelated for a purpose, which for a company is to achieve the objectives of the management. The management of a company has overall corporate objectives, usually starting with an overall objective such as to be sufficiently profitable and to still be in business in five years time. To achieve these, the immediate objective is to complete a successful ship project.

All of the necessary control as described above is managed through information. The information must be of good quality so that the management has an accurate basis on which to take decisions about changes to the sub-systems. The timing of the information is important because it must be available in sufficient time for effective action to be taken. The information must have relevance so it deals with meaningful issues, such as deviations and does not offer unimportant trivia.

The presentation of the information is very important because it must be readily understandable to the recipient. There must be appropriate detail, so that the recipient has enough, but not too much so as to make the issue obscure. There must also be an appropriate volume of information, enough to inform decisions and no more. Extracting meaning from a lot of figures is very difficult if there is limited time to study them. Having information about many activities which are correct can distract a user from the few deviations which it is important to correct. Graphical presentations can be useful, but users may need specific training in their correct interpretation.

Once the data has been collected about whatever situation is to be studied, and processed into useful information as above, it then has to be communicated to the users. There is a necessary sequence of actions to carry a piece of information from its source to a manager (for example) who needs to act. The model in Fig. 18.1 outlines this sequence.

The system can be represented simply as a process, the sequence of activities in Fig. 18.1, which take the information from source to a resulting action. A system can in theory be closed, so that it has no interaction with the environment, but this will not happen in practical terms in a working environment. Real systems are open in that they interact with their environment.

The effect of the environment is identified as "noise", that is extraneous information, or any disturbances which may distort the message during transmission. Also for long term survival real systems must adapt to environmental changes, which can be alterations to the operations in the shipyard. At any point in the model above, there is some potential for error. The source of error may be some data which is processed, but that process is flawed, or a piece of information may be read and misunderstood.

The source is the point at which an event occurs, a need is identified or any other happening about which someone needs to know. Typically it might be the completion

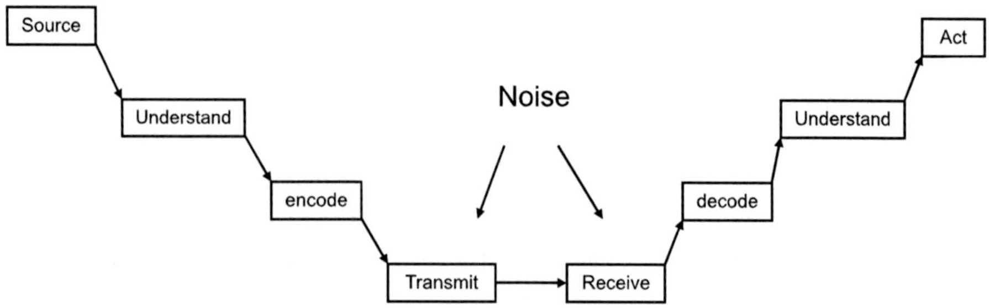

Fig. 18.1 Communicating information

of a piece of work which must be reported to a manger. The first potential problem with the communication is that the initial happening may be incorrect, so the piece of work is not really finished. The event also has to be understood correctly, so the correct information is found.

The information is then encoded so as to be in a form that can be sent to a recipient. It may be as an email, phone call, text, input to a computer system or simply spoken. Any of these activities offers a further opportunity for error. For example a work activity code may be wrongly input to a computer.

The encoded message then goes to a transmitter, which may be the human voice, telephone, radio or an input to other device. The channel is the conduit from the transmitter to the receiver. Again, errors may occur. The human voice will have difficulties in a noisy working environment or radio communications may be distorted. As a result the message may be distorted before it reaches the receiver.

The receiver may fail to receive the correct message. Mishearing in a noisy workplace or on a poor radio link, or through poor handwriting, the information which should reach the recipient does not do so. As a result the message is not correctly decoded, so the wrong action is decided on and taken. Also the information can again be misunderstood and so the action is again incorrect.

Clearly in practice there are safeguards, systems are carefully constructed to avoid the problems described and wrong information is likely to be questioned and then clarified. A computer system should be designed to query an input that is outside normal limits. But, given the sheer quantity of information in a complex environment such as a shipyard, and particularly where there is some pressure on the people because of urgency or the need to correct problems, wrong information will sometimes be found.

In a large, complex system such as a shipyard there are usually further sub-systems, each with their own information inputs and outputs. Each sub-system deals with part of the external or internal flow of information. The management information system which is adopted for the management of the organisation will support the processes or activities.

An important element in an information system is the feedback loop, which basically evaluates outcomes, typically outputs from processes, against requirements from those outputs. Deviations are then investigated, essentially in terms of the risk

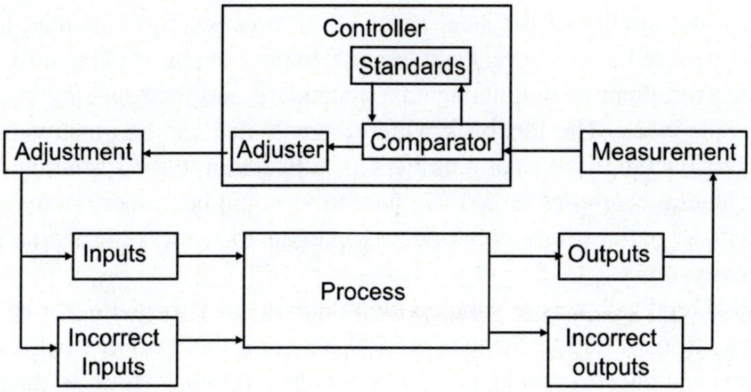

Fig. 18.2 Outline feedback system

they pose to the project. Most deviations should be small, relatively simple to correct and containable within the work station where they occur.

If this is the case then an adjustment to the inputs to the system will be made and further feedback will confirm that the problem and associated risk has been eliminated. This can be through better instructions, small machine adjustments or enhanced supervision.

The process model in Fig. 18.2 shows the basic feedback loop. The process converts inputs, for example raw materials into outputs, in this case parts for later production processes. As an example a piece of steel is cut to create one or more parts. As well as steel, the inputs to the process will be workers, equipment for measuring and cutting. The parts are reviewed, usually sampled for large quantities or individually checked for large assemblies. The chapter on quality management outlines the processes used for this. The outputs are checked against the requirements and if there is a problem then an adjustment within the control mechanism will make changes which are seen as necessary.

There are undesirable inputs and outputs to processes, ideally very few. Examples would be damaged or inaccurate input materials, and also damaged or inaccurate parts being produced. Occasional serious problems will occur and can generally be fixed as one off occurrences.

For small output deviations which occur regularly, even after some adjustment is made to the inputs, then further action will be required. This may require action to examine the actual process in use to ensure it is operating correctly. Where this is not the case then action is required to alter the actual process so that it will produce the outputs which are as requirements.

As part of the management of quality, the part, or often a sample of many parts are checked against the production requirements. In the case of steel parts the critical characteristics to be checked are dimensions. If these are correct then all is well.

However if the dimensions are found to be out of tolerance then some action is required. The adjuster will act on the process inputs to make a correction. This could be to retrain workers, renew equipment, to calibrate measuring equipment or to improve the quality of the steel input to the process. Systems tend to fluctuate about the expected output levels and this is a feature of statistical quality control.

The measurement of outputs against a standard and then making corrections is known as feedback. The feedback which is generated can be positive or negative. Positive feedback reinforces or amplifies an adjustment to the system under consideration. General examples of positive feedback would be power assisted braking in a motor car, or perhaps linking marketing budgets to success in specific markets as a business example.

Negative feedback acts to dampen the fluctuations. This makes for an inherently stable system. An example is stock control, where if the level of stock is high, then reordering is reduced or delayed to correct this. If more stock is then used, the feedback will point to increasing the orders so the stock level tends to return to a desired level.

The timing of feedback information is important. If the outputs from the system are oscillating, feedback to correct this must be timely, or the effect may be to increase the fluctuations. Negative feedback may become positive in a system with long delays between the deviation being identified and the feedback being used to make a correction.

The need to structure information to make it relevant and clear has been mentioned. This is to try to ensure that problems are identified quickly for action. The law of requisite variety is a basic concept in cybernetics, the study of communications and control in animals, machines or organisations. Large, complex organisations have high variety, which is simply the number of states the system can have. The number is a function of the numbers of people, with different objectives and motivations. It is also a function of the information generated by those people and fed back to the management.

The management of such large, complex systems must have equal variety, so that the information reaching the management for control purposes can be managed. The requisite variety is the variety needed by management to match the system. In real life, many basic control mechanisms, like budgets, are inadequate for the purpose.

To manage the organisation, two basic strategies can be adopted. The first is to attenuate the variety to be managed. Obvious means of doing so are to create rules for staff behaviour, such as timekeeping, safety, waste management and production targets. Most people will follow these so only deviations need to be managed. In terms of information flow, management should limit the requirements again to deviations. It is not necessary to be told what is on target, only those aspects which are not.

The second strategy is to amplify the ability of management to manage, so the management variety is increased. At its simplest this creates a hierarchy, so the CEO only manages directly a small group of executives, who then each manage a group in their functional area of responsibility. That group in turn has lower level managers, and then supervisors who manage a small group of workers. This limits the quantity of information (the variety) each person has to control.

The information is filtered so that most decision making, usually about deviations, can be managed locally, with anything affecting a wider group being transferred up the hierarchy for a solution. Although this works as a guiding principle, human behaviour is such that some workers, supervisors and managers may try to conceal problems. In particular behaviour which transgresses the rules and may lead to sanctions is likely to be concealed.

The point has already been made that if a problem, for example a conflict between two project managers about priorities, has to be escalated to higher levels of management, there is a danger of time being lost. Meetings have to be set up, the decision may have to go higher and when made has to be accepted and implemented by both groups. As a result when the situation arises, there is a need for rapid action and that requires access to the information needed by management. Information that senior managers would not usually require may be key to the problem solution.

This is part of the need for a management information system in the shipyard which can support the management of the projects being carried out. Over the life of a project, and for past projects, the management may need to have instant access to the project status from initial enquiry to final invoice payment. Reliable information is essential for managers and supervisors at all levels in the company. Good quality decision support data for senior managers is critical, when it is required.

The management information system should link all the activities on the activity map. As a result there will be a complete set of information about the projects in a company. This information should be organized so that individuals are supplied only with the information they require to manage their specific tasks. More senior staff, for example the project manager for a ship, should have access to summary information, though some ability to override limitations and see more may be required on occasion. A danger to avoid is senior managers continuously trying to view all the information that is available and trying to micro-manage individual projects. There has to be trust in the staff generally.

There are a number of products available for management of information, some specifically designed and developed for the shipbuilding industry. Most have their origins in one aspect of the business, for example planning of projects, accounting or design and have been extended.

Any system which is adopted should be comprehensive, but should also be able to link into existing software in use in the company. There is no merit in disturbing a well established and fully operational system, to introduce something new and unfamiliar. Any new system will have been influenced by the requirements and attitudes of the original users, and the subsequent clients of the software vendor, so will not necessarily be what potential users expect. It is difficult enough to introduce a new information technology system in any case, so the least disruption to what exists is important. A new system should support the working methods of a company, not force unnecessary changes.

An outline of a comprehensive management information system is set out here.

For marketing, the process of finding good enquiries, it is necessary to be able to easily track the histories, people and ships within a company's trading community. Having better knowledge of the particular market allows the staff to focus their marketing efforts and improve the results. Immediate access to past, current and potential customers, and their status will inform marketing contacts, visits and other activities. This offers feedback from prior work which can support staff to focus their effort on the likely most productive areas of the market.

For initial project development, leading to discussions on price and then tenders, the key is to develop realistic cost estimates. The mechanics of estimating are actually relatively straightforward, but to be able to generate more accurate estimates is helped by immediate access to previous estimates, tenders and actual costs. A database of past and current estimating information is essential. Most shipyards have a well developed existing estimate format, and the ability to use these in parallel with an information system will make the work easier and avoid any possible loss of specialist information.

All of the above is of great assistance in negotiation with potential clients to try and obtain profitable, low-risk contracts. Negotiations can be better informed with full tender, client and contract histories. It is also important to be able to test 'what if' commercial scenarios to win business, and keep a record of different likely outcomes and customer reactions.

Purchasing has become an increasingly important activity in shipyards. There is much more information to manage, with the objective of speeding up the purchasing activity while maintaining full control. Better communication between purchasing, technical, planning and production departments will make the scheduling of purchasing work more realistic. A complete system will also allow for tracking of the full purchase cycle from an initial material requisition to use in production and testing. Further, expediting and managing fast-track purchases where necessary can be improved. Greater accessible information on suppliers is valuable, on aspects such as their reliability, product quality and service.

Once a project is in production, the key is keeping control of operations. This requires the ability to monitor contracts in real time and be automatically warned of problems. Production supervisors and management need to have the information immediately at hand. Administrative procedures must be automated and simplified so managers can get on with managing rather than staring at a screen. At this point the clear presentation of information becomes critical.

All work packages will be entered to the system, with links to materials required and then to purchasing status for these. All work instruction requirements will also need to be accessible so that availability can be checked in advance of start dates. Work package status will be either not started, in progress or complete. Only in progress packages should be able to have man-hours recorded to them, and any apparent deviation from what is expected must be readily identifiable.

Labour cost recording is important as part of the progress monitoring. The most useful feature here for an information system is a capability for rapid data entry and checking. Even if the data collection is automated in some way, for example using bar codes or other identification to locate workers, checking is needed. As jobs progress, recorded labour hours are entered into the system, which must maintain both daily and total hours. The system will also have to maintain all necessary personnel information, including training received, other qualifications and personal data.

All quality management actions and results will be recorded, so that work packages are known to be correct and system tests completed.

Accurate costs for all work will be recorded as the project progresses. There should be a running total from day one of the contract and as costs are recorded that total should be automatically recalculated. There should also be potential for authorized manual input to the process to allow commercial decision-making. All the above applies to any work scope changes as agreed with the customer during the project.

The benefit of a fully integrated system, which can be customised to fit company requirements and also interface with any existing systems is that once any information has been input it is immediately available in the right format needed by each authorised user in the company. Consistency will ensure that all the information that personnel need to do their job is brought together in a single place and presented at considered and appropriate levels of detail. The ability to access past work records as easily as current ones allows comparisons to be made and acts as the memory for the organisation.

The ability to create reports for specific needs using any of the data held in a system is important, for progress meetings, both internal and with third parties including the customer. Reports should be exportable in a recognized format so they can be shared as needed with third parties.

An effective management information system is essential to support management of complex ship projects.

Chapter 19
Completion and Evaluation

Done is better than perfect
—Sheryl Sandberg
Your most unhappy customers are your greatest source of learning
—Bill Gates

After the completion of any project, including a ship, the final activity should be to make an evaluation of how the work was conducted. The project will continue to the end of the warranty period after delivery, during which any issues with the owner will be resolved and any defects remedied. The evaluation of the shipyard work though can be largely carried out at the time of delivery to provide feedback to future projects.

Ultimately the success of a shipbuilding project is determined by whether the financial targets for the ship construction were met, and ideally a profit made. However from the view point of the shipyard, it is also important to determine whether the ship was completed to timescale as defined in the baseline plan, whether the completion was within the budget and whether the correct quality was achieved efficiently. A number of questions can be asked as a basis for the comparison of the initial project management plan with the actual outcome. The project outcome can also be reviewed from the perspectives of the different stakeholders. Some projects are not good for perhaps a contractor, but others engaged in the same project regard the outcome as very satisfactory.

A project may have been completed within the budget set, but this should not simply be accepted at face value. The discussion of earned value demonstrates that the budget should include contingencies and reserves. If these were used there still may have been opportunities which were missed to save costs. The budget structure set up at the start should be reviewed to see to what extent the performance target was achieved, or the management reserve used. Even if they were not used a more detailed review of the expenditures may show opportunities for future costs savings.

If the project went over budget then there is likely to be more urgency in assessing where and why the cost overruns occurred. Ideally potential overruns should have been identified much earlier during the project monitoring process and the lessons found should have been applied.

G. Bruce, *Shipbuilding Management*,
https://doi.org/10.1007/978-981-15-8975-1_19

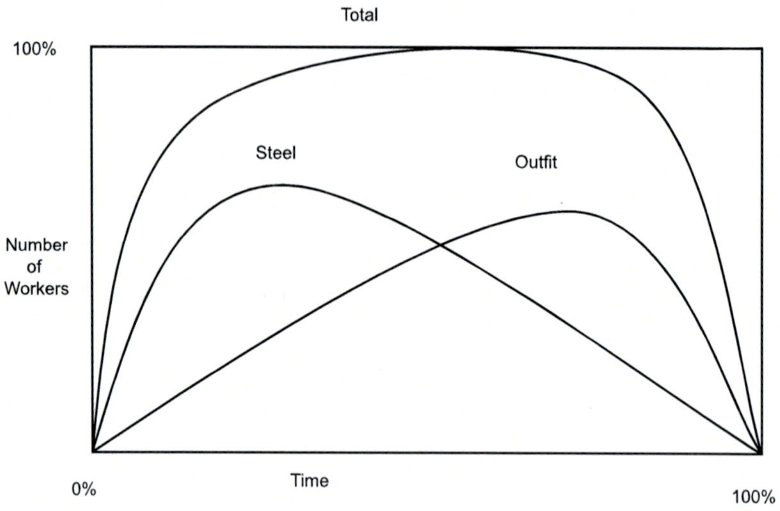

Fig. 19.1 Labour utilisation for a project

There may be a strong temptation to see a completed ship as a final stage of the contract, making due allowance for the warranty period. This is a mistake. A commercially successful project may conceal work completion later than planned, provided the contract was such that the lateness was acceptable. If the lateness was due to some shipyard problem, this should be fully understood and plans made to avoid the same in the future. If the lateness was down to some acceptable, external factor then this should be evaluated as a risk which occurred, and steps taken to avoid the same risk in future. Even if the work was all to plan, opportunities to save time may have been missed and future planning for ships can take this into account in order to save time in the future.

Figure 19.1 is an example of the information which can be taken from an evaluation and then used for future project planning. The figure shows the weekly total number of workers engaged on the project over its duration. The number rises then falls during the project, and is the same information as in an S curve. If the data is normalised then the curve can be used to predict the numbers of workers who will be required for a new project. Applying the estimated total hours and the timescale to the curve will allow weekly totals to be estimated.

The figure also shows the total divided into steel and outfit workers, which can give further information, and a more detailed division can be made. An unplanned increase in resources near the end of the project will show and be a subject for investigation. This provides a good starting point for a new project, as an accurate baseline, even when the expected ratio of different disciplines is likely to change.

The quality of the project work is determined largely by acceptance of the ship on behalf of the customer and the relevant regulators and inspectors. Beyond the major issues, such as the speed and capacity of the ship, and the safety requirements, there are many smaller aspects which might result in discussion and potentially argument.

The technical requirements for the ship should have been set out in the specification for the ship, and thus formed part of the contract. Any actual or potential problems will usually be identified and fixed during testing and trials of the ship.

However when discussing quality, it was made clear that finding and then fixing quality problems will result in a satisfactory ship, but at extra cost and probably time. The ability to deliver a complex ship which operates as specified should never be confused with the ability to run a successful project. The technology of the product can conceal failings in the technology of the shipyard. So it is necessary to review quality issues on a project and identify how changes in the processes used in the shipyard can be amended to prevent future occurrences.

The evaluation is not confined to the end phase of the shipbuilding project. Basically, as the work progresses, the shipyard should identify any good or bad aspects and record them for future reference. At a detailed level, in workstations, it can be possible to find and implement improvements during the progress of a specific project.

Repeating mistakes or failing to take up good ideas are depressingly common happenings. Avoiding these errors is essentially part of good quality assurance procedures. The shipyard uses the evaluation process to identify ways in which the future management of similar projects, and the on-going management of specific activities in the shipyard, might be improved for the organisation.

It is also important to evaluate the way in which the project was managed, again as a basis for improvements in the future. Sadly in some cases the evaluation leads to dismissal of a project manager, which is not a good way to promote open evaluation or learn lessons for the future. Having reviewed a project critically, it should be possible to identify how improvements might be made in the future.

The project review will basically cover all the work done in the design, work preparation, purchasing and production to deliver the ship. The focus will be first on what went wrong, rather than any congratulations on what went right. It is the errors that result in cost and time overruns. Then the review can move onto what went well but could have been better. Congratulations should come last. There is a problem here, in that by the time work is completed and the data for evaluation has been obtained, the people responsible will usually have other concerns about new projects.

Quite correctly, those responsible for completing their part of the of the ship construction project will focus their efforts on their current and next tasks. That is unless there is an underlying problem in a work station which does need urgent attention. So it is necessary to have procedures to collect information on what happens during the project and to retain that information so it is made available as input to the planning and preparation of future projects. Basically, that is when the people will have their attention on those aspects of the new project where the potential for improvement has been found.

The final stage in a project is to complete the work and agree closure with all the parties to it. Reaching and agreeing completion can be sources of disputes. Despite this, the issue of deciding what is complete should not really cause any problems for the construction of a ship. The specification and contract will have defined what is to be produced and the delivery to the owner will be agreed on that basis. There may well be changes to the specification during the life of the project, hopefully carefully managed, but the completeness should still be easily agreed.

Prior to delivery of the ship, the following are normally checked and confirmed as being correct. All the instruction manuals for operation and maintenance of the ship should have been received by the owner. The planned maintenance program must have been initiated, all owner's supplied items, consumables and other operating requirements have been received and stored safely. Finally any necessary items for a delivery ceremony must be readily available, for example flags, emblems and uniforms.

Depending on the complexity of the ship, some crew may be standing by the ship for a considerable period. This is particularly the case for military ships, where there may be considerable quantities of stores and other items to be looked after by the shipyard prior to handover of the ship to the owner.

Delivery of the ship can be made after completion of all the site inspection and construction supervision. The chief inspector will ensure that all approved plans have been evaluated and corrected in compliance with the ship as-built. Once delivery of ship is accepted and the authorised person has signed the necessary protocol on behalf of the owner, a number of documents must be submitted for any financial institutions and for the flag state administration.

Unless there is a major failure to meet the specification, for example if the speed is inadequate or the cargo capacity is below that which was specified, then the ship will be accepted as meeting the owner requirements. If there is a major failure then the shipyard has not managed the project correctly at some point. If the owner will take delivery of the ship after deduction of whatever penalty was specified in the contract then the shipyard will have to accept this. Also if the owner enforces a right to reject the ship then the only recourse is to return the payments made and perhaps try to re-sell the ship. The shipyard might be able to pursue a sub-contractor or supplier who has caused the failure, but will not be able to have any compensation for the whole consequential loss.

Even if there is no major failure, there may still be some dispute, for example over the quality of the ship. If there is such a dispute this is when the initial specification and the related contract really matter. If the specification was complete and detailed, then confirming that the ship complies with it should be easy. If the contract is unambiguous, for example with clear references to the appropriate standards which apply and the shipyard has complied, then acceptance of the ship should be straightforward. However anything left vague in either contract or specification will be a source of potential conflict.

Practically, no ship is going to be perfect and as the discussion earlier in the book mentioned, where defects or other problems are identified the actions to be taken must be agreed. Again, these should be covered by the contract conditions, so that the actions are agreed quickly and easily. In some cases a shipowner may want to delay completion because of his cash flow or a poor market for shipping. The issue of distortion of steelwork is one which has been mentioned earlier and sometimes it may be difficult to agree when the structure is in a suitable condition.

Assuming that all parties are being reasonable, then decisions have to be made about closing the contract in an acceptable way. If the project has been managed well, then the earlier inspection delivery reports for nearly-complete work should have picked up any potential problems in good time. All the progress monitoring and inspections of work in progress were designed to find problems and their solutions to ensure there would be no impact will occur on the construction schedule or the eventual performance of the ship. All the tests and inspections will have been documented.

In closing out the contract, some technical requirements are for the owner to receive certified completion of scope of work, including variations, performance analysis after a period of operation and performance improvement initiatives during the warranty period. All auditable documents must be agreed and signed, technical information complete and passed over, including operating instructions, as-built drawings and test records. What triggers the warranty period and when is agreed, residual materials are transferred to stock or disposed of and customer-owned property is accounted for and properly disposed of.

Commercial questions to check are whether all bills have been paid, have all outstanding disputes been handled satisfactorily, are the terms and conditions of the warranty understood, is the close out report agreed and are residual bonds and credits cancelled. Any claims and disputes must be resolved or recorded, potential claims and disputes eliminated, fees or royalties due under licence agreements paid, internal order numbers closed so that no further costs are allocated to the project and obligations with respect to offset arrangements fulfilled.

The owner's project manager will have clear acceptance arrangements for inspections, tests, trials and reports. Final tests and trials often require a surprising amount of time and effort to prepare for and carry out. Any deficiencies that prevent tests and trials from proceeding will have been corrected, possibly only as short-term, expedient quick-fixes. In this case there is residual work to be done so that the proper solution is implemented later. Potentially this will be after delivery within the warranty period, provided the ship is safe to sail. These are outstanding inspection reports agreed not to be closed out, but for which a price adjustment will be made. Also inspection reports that the shipyard agrees to correct within the first half of the warranty period, and against which the owner has withheld some payment. Inspection reports in dispute where some payment may be withheld against these items subject to post-delivery negotiation or arbitration or litigation are listed.

There will also be a review of non-hardware deliverables prepared by the shipyard. These include as-built drawings which will show the final arrangements and confirm design changes are as agreed. All certificates, operating manuals and other documents required by international or national regulations will also be confirmed.

The project manager will have to approve spare parts lists and arrange for their timely arrival. Another need is to develop documentation for delivery, including three lists.

Management will identify warranty items according to contract procedures. Warranty items should not be confused with outstanding deficiencies. The contract will have set out the procedures, rights and responsibilities of each party in respect of the timing, notifications, correction of warranty defects, costs and withheld sums. The burden is on the owner to prove that there is a valid warranty matter and is if there is then it must be managed in a timely way.

A guarantee period is part of the contract, and this is provided for as a means of ensuring any small problems with the ship which do not prevent its acceptance by the owner are corrected. The owner will usually have an immediate need for the ship and will not delay acceptance provided deficient items are fixed. The guarantee period also provides for the correction of any problems which occur once the ship is in service. These are usually minor, and can again be fixed quickly, by shipyard staff or an appointed shiprepair company.

Occasionally large scale problems occur, which can be a problem because if the owner cannot operate the ship then there will be a loss of money and the owner will want the shipyard to pay. The shipyard will usually then wish to pass the solution of the problem on to a sub-contractor or the supplier of the relevant equipment. Such problems are generally equipment related, and resolution can result in argument and sometimes legal action.

There is potentially a positive note to the guarantee period. If a purchased item works properly, then this will not usually cause much comment, after all it is what was expected by the owner. If there is a problem which leads to distrust of the supplier shipyard, or even to a legal argument, then the owner is likely to make a lot of comments, for example if asked about the reliability of the shipyard. If however the ship develops a problem and this is fixed quickly and effectively by the shipyard, this is likely to result in positive comments. Shipyards which have been recently developed, and employ a labour force which is new to industry, are likely to make mistakes and have post delivery quality problems.

This was a feature of some new Asian shipyards in past decades. The reaction was to accept these problems as likely and then to provide a serious after sales service, with guarantee engineers aboard the ship as it operated, tasked with problem solving. South Korea in particular developed this procedure into a well known and appreciated service and gained a good reputation, in addition to offering ships at low prices which was always the main incentive for owners.

This neatly returns to the initial activity described in this book, the importance of the marketing of the shipyard. It is generally recognized in marketing any product that a current or recent customer who needs more of that product is the easiest opportunity for more work for the supplier. So the provision of reliable ships, but also and the rapid solving of problems, is a very effective marketing tool. As the first requirement of shipbuilding management is to obtain orders for ships to build, this emphasizes the importance of carful closure of a ship construction project, evaluation of the production process for each ship built and actions to improve for the future.

Further Reading

It is worth mentioning a small number of books and other sources for further reading. There are very many alternatives to the very small selection here and specific information can readily be found through a web search. Most of the large number of books on the subject focus more on the ship than the construction process, though some do describe the actual production of parts, assemblies and construction for ships. Some shipbuilding companies have useful information on their web sites, in greater or lesser detail, with numerous illustrations.

Other sources include professional organisation such as the Royal Institution of Naval Architects, (RINA) in the UK and the Society of Naval Architects and Marine Engineers (SNAME) in the USA. A textbook on Ship Design and Production is available from SNAME and both publish relevant papers and conference proceedings.

Various United Nations agencies offer an international perspective, including the OECD which reports on the market and other industry issues including environmental issues. Information on the current and recent state of the international industry is also available from some national shipbuilders' associations, particularly Japan.

Maritime Economics, by Martin Stopford, is a comprehensive look at the whole subject for specialists. Competitiveness has been the subject of numerous studies over recent decades by national and international agencies, including the UN, OECD and the European Union. There are several commercial and consulting organisations who have completed studies commissioned by government and other organisations.

There are many good general introductions to industrial management techniques, with a lot of information on work study, planning and other topics. Any book on production and operations management will explain the many techniques, including those outlined in this book. The Handbook of Project Management, by Trevor Young, is one very good introduction to the subject and a helpful practical guide.

Many organisations, commercial and professional can offer further information, including on estimating, planning, progress management and project completion. Two are, the Project Management Institute and the Association for Project Management in the UK. They also offer material on quality management and control. Quality is a major theme in modern management and many commercial offerings

G. Bruce, *Shipbuilding Management*,
https://doi.org/10.1007/978-981-15-8975-1

can be found. A useful starting point for information is the International Standards Organisation.

Commercial shipbuilding contract templates are available from, among others, Baltic and International Maritime Council (BIMCO) and the Shipbuilders Association of Japan.

Numerous books on materials management are available, though some deal with mass production industry and supply chains, rather than the specialised materials and components which are found in shipbuilding.

The Chartered institute of personnel and Development in the UK and other national bodies have substantial information on human resources. Health, safety and environmental management can be found on national government websites.